# Science Fair Projects with
# ELECTRICITY &
# ELECTRONICS

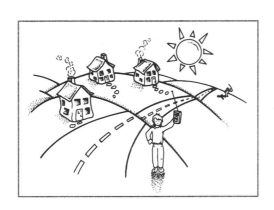

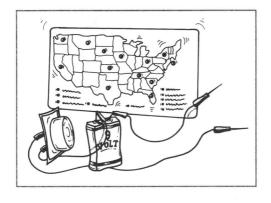

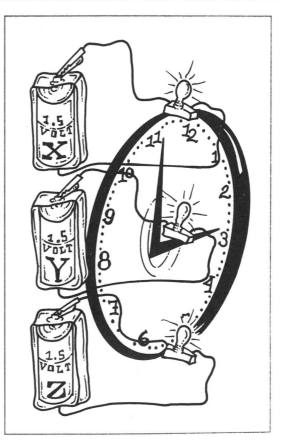

## Bob Bonnet & Dan Keen
### Illustrated by Karen McKee

STERLING PUBLISHING CO., INC. NEW YORK

Edited by Claire Bazinet

**Library of Congress Cataloging-in-Publication Data**

Bonnet, Robert L.
  Science fair projects with electricity & electronics / Bob Bonnet & Dan Keen;
illustrated by Karen McKee.
  p. cm.
  Includes index.
  Summary: Nearly fifty projects on electricity and electronics, designed for
science fair competition.
  ISBN 0-8069-1300-2
  1. Electricity—Experiments—Juvenile literature. 2. Electronics—Experiments—
Juvenile literature. 3. Science projects—Juvenile literature. [1. Electricity—Experiments.
2. Electronics—Experiments.  3. Experiments.  4. Science projects.]  I. Keen, Dan.
II. McKee, Karen, ill.  III. Title.
QC527.2.B66  1996
507.8—dc20

95-51492
CIP
AC

10 9 8 7 6

First paperback edition published in 1996 by
Sterling Publishing Company, Inc.
387 Park Avenue South, New York, N.Y. 10016
© 1996 by Bob Bonnet & Dan Keen
Distributed in Canada by Sterling Publishing
% Canadian Manda Group, One Atlantic Avenue, Suite 105
Toronto, Ontario, Canada M6K 3E7
Distributed in Great Britain and Europe by Cassell PLC
Wellington House, 125 Strand, London WC2R 0BB, England
Distributed in Australia by Capricorn Link (Australia) Pty Ltd.
P.O. Box 704, Windsor, NSW 2756 Australia
*Printed in China*
*All rights reserved*

Sterling ISBN 0-8069-1300-2 Trade
     0-8069-1301-0 Paper

# CONTENTS

# CONTENTS

# INTRODUCTION

"I listen and I forget. I see and I remember. I do and I understand." —Unknown

Welcome to the exciting exploration of the world around us...the world of science. Our environment provides us with many things to observe and processes to understand. Knowledge is gained by observing and questioning.

Science should be enjoyable, interesting, and thought-provoking. That is the concept the writers wish to convey. While this book presents many scientific ideas and learning techniques that are valuable and useful, the approach is designed to entice the student with the excitement and enjoyment of scientific investigation.

The scientific concepts introduced here will help the student to understand more advanced scientific principles. Projects will develop science skills needed in our ever-increasingly complex society: classifying objects, making measured observations, thinking clearly, and accurately recording data. Values are dealt with in a general way. Respect for life should be fundamental. One should never harm any living thing just for the sake of it. Disruption of natural processes should not occur thoughtlessly and unnecessarily. Interference with ecological systems should always be avoided.

The activities presented in this book target sixth through ninth grade students. Most of the materials needed to do the activities are commonly found around the home or are easily available at minimal cost.

Safety is and must always be the first consideration. We recommend that all activities be done under adult supervision. Seemingly harmless objects can become a hazard under certain circumstances. Even a bowling ball can be a danger if it is allowed to fall on a child's foot. All of the projects in this book work with low voltage to insure safety.

Science projects motivate students to learn, develop thinking skills, promote a questioning atmosphere, and teach problem solving. These are some of the many benefits that can be gained from doing science projects. Spin-off interests can develop. In doing a science project about weather, while using a computer to record weather data, a student may discover an interest in computer programming.

The authors recommend parents take an active interest in their child's science project. In addition to safety, when a parent is involved, contact time between the parent and child increases. Such quality time strengthens relationships, as well as the child's self-esteem. Working on a project is an experience that can be shared. An involved parent is telling the child that he or she believes education is important. Parents should support the academic learning process at least as much as they support Little League, music lessons, or any other growth activity.

Adults can be an invaluable resource for information, one that the student draws upon. Older people can share their own life experiences, while they provide transportation, taking the child to a library or other places for research. In our school, one student was doing a project on insects so his parents took him to the Mosquito Commission Laboratory, where he had an opportunity to talk to entomologists.

Science is all around us. It is important to us. If affects the food we grow, how we transport it, how we eat it. Science and its attending technologies have enabled us to build telephones, spaceships, and microwave ovens. Science is in the news.

# YOUR SCIENCE PROJECT

Science is the process of finding out. "The scientific method" is a procedure used in science fairs, consisting of several steps: state the problem, make a hypothesis, set up an experiment and collect information, record the results, and come to a conclusion about the hypothesis.

A science project starts by identifying a problem. The statement of the problem defines the boundaries of the investigation. For example, air pollution is a problem, but you must set the limits of your project. It is unlikely you have access to an electron microscope, so an air pollution project could not check pollen in the air. This project might be limited to the accumulation of dust and other visible materials.

Once the problem is defined, a hypothesis, an educated guess about the results, must be formed. Hypothesize that it is dustier in a room that has thick carpeting than in a room that has a hardwood or linoleum floor.

Set up an experiment to test your hypothesis. Smear petroleum jelly in two cereal bowls. Place one on the floor of a room that has the carpet. Place the other on the floor in a room that has a hardwood or a linoleum floor. After two weeks, look at each bowl. Use a magnifying glass. Write down the results of your experiment. Reach a conclusion as to whether or not the hypothesis is correct. To learn, an experiment need only be procedurally correct. The hypothesis doesn't have to be proven correct in order for the project to be a success. Knowledge is still gained.

Science projects must make "assumptions." Certain things are assumed. In our example above, we are assuming that all rooms with carpets are similar and all rooms with hardwood or linoleum floors are similar. Therefore, if the project is duplicated by someone else in the house, or elsewhere, they will get the same results.

When doing some projects, the number of test items must be considered. This is called "sample size." If we hypothesized that 4 out of 5 people like a certain brand of soda, would you trust the results of that survey if only 5 people were interviewed? The results would be more trustworthy if 500 people were asked. If a project is to germinate a bean seed under certain conditions, using only one seed would make the results untrustworthy, because seeds often won't germinate under any conditions. A larger number of seeds (a larger sample size) is needed for accurate results.

Collections, models, and demonstrations usually do not follow the scientific method and are not allowed in science fair competitions. You could hypothesize that a model airplane will fly when you put it together, but that is not very scientific because it is supposed to fly when it is built correctly. You can turn a collection into a good science project if you use it to do comparisons. A collection of leaves from the local area could be a reason to hypothesize that the leaves of broad-leafed trees are similar in structure to broad-leafed house plants.

Be sure you know the rules of your science fair competition before you begin a project. Does the competition allow collections? Is special permission needed if you use a live animal in your project? Can more than one person be a part of a project? Real science is done in groups, but usually the top scientist gets the credit. An adult may help supervise, get materials, provide transportation, check for safety, and offer a second pair of hands if needed, but unless the science fair competition allows groups, you will most likely be expected to do the actual project work on your own.

Safety is the most important consideration. If you can't do a project safely, then don't do it. For example, you can't deal with bacteria that could infect a population.

Ethical rules must be followed also. It is unethical to hypothesize that one race or religion is better than another. Your project cannot be inhumane to animals.

When choosing a science fair topic, pick something that is interesting to you, that you would like to work on. Then all of your research and study time is spent on a subject you enjoy.

# ELECTRICITY & ELECTRONICS

This book deals with a wide variety of topics related to electricity and electronics. Choose from a subject area that you like best. We suggest making a "schematic diagram" of projects that use electrical circuits. This will enhance your display visually and increase understanding. A schematic diagram is a pictorial layout of an electrical circuit, showing the arrangement of components and how they are connected. Symbols are used to represent the various components. Some standard symbols are shown in the Appendix.

No one knows what electricity is. We have learned how to generate it and how to make it do work. We understand its properties, how it behaves, and what takes place at the atomic level. But we really don't know what it *is*.

Electricity refers to the movement of electrons through a conducting medium (a pathway) such as copper and silver. Electron movement is energy that we can put to work to improve our daily lives.

The difference between what we call electricity and electronics is small. Generally, electronics is considered to be a branch within the science of electricity. Electronics deals with electricity as it moves through and is affected by certain devices. These devices include resistors, capacitors, coils, transistors, and integrated circuits.

The human race has harnessed electric power to perform tasks from illuminating a simple light bulb to making a calculating machine called a computer which can do math operations at tremendous speeds.

Electronics is among the most rapidly changing sciences. Technological advancements are made every year. Consider how far computer technology has come in the relatively short time since the first commercially available home computers were introduced around 1977.

In most homes today you will find a wealth of electronic marvels: computers, stereos, televisions, radios, telephones, laser music discs, copy machines. Electrical appliances have enhanced our standard of living: washing machines, hair dryers, vacuum cleaners, toasters, microwave ovens, electric can openers, refrigerators.

Many specific topics fall under the blanket category "electricity and electronics." Topics in this book include electromagnetic forces, static electricity, current flow, motors and generators, resistance and capacitance, generating electricity, solid state electronics, and radio frequency energy. You have only to make your selection, and begin your project.

# SECTION ONE
# ELECTROMAGNETIC FORCES

Magnetism is an electromagnetic force that occurs between certain materials in nature. Magnetism can be found in rocks called "lodestones" (sometimes referred to as "magnetite"). Lodestones exert an attractive force on materials containing iron. Iron can be turned into a magnet by stroking it with a lodestone. It can be said that we inhabit a magnet. Early experimenters, around 1200 A.D., determined that the planet we live on is a huge magnet.

Electricity and magnetism are related. It was discovered that, when an electric current flows in a wire, the needle of a compass placed next to the wire will be deflected. The same holds true in reverse, namely that, when a magnetic field is moved past an electrical conductor, current is generated. This concept is the basis for electric motors and generators.

# PROJECT 1–1
# Attractive Force
## *Relationship of wire turns to attraction in electromagnet*

---

**YOU NEED**

- large iron nail
- hookup wire (18 or 20 gauge, solid core)
- two 6-volt lantern batteries
- iron filings
- index card
- three insulated jumper leads, with alligator clips on each end
- wire cutters
- gram-weight scale
- two sheets of plain paper

---

**W**hen current passes through a wire, a magnetic field is built up around the outside of the wire. You can make a simple electromagnet (a magnet that attracts only when it is connected to a power source) by wrapping wire around a nail and connecting one end of the wire to the positive side of a battery terminal and the other to the negative. Hypothesize that the magnetic field will be stronger if the voltage is higher.

Cut a length of hookup wire, about 18 or 20 gauge, to a length of about three to four feet. Wrap the wire in a spiral motion around a large iron nail. Strip 1 inch of insulation off each end of the wire to expose the bare wire inside. Using an insulated jumper lead with alligator clips on each end, attach an alligator clip to one of the ends of the wire. Connect the other alligator clip to the positive terminal of a 6-volt lantern battery. Using another insulated jumper lead with alligator clips, attach an alligator clip to the other end of the wire that is wrapped around the nail. Connect the other alligator clip to the negative terminal on the battery.

NOTE: Since the wire that is wrapped around the nail is almost a dead short, having very little resistance to current flow, the battery will not last very long before it begins to lose its power. For that reason, only connect the alligator clips to the battery while the experiment is actually being done. One alligator clip can be connected at all times, since no current will flow unless both are connected.

To prove that the hypothesis is correct, the strength of the electromagnet must be measured. This can be done by seeing how many iron filings the electromagnet can collect. The iron filings that are collected can be weighed.

Pour some iron filings onto a sheet of typing paper. To keep the electromagnet from getting dirty, rest an index card on top of the iron filing pile. Hold the point of the nail on top of the index card and connect the battery. Slowly pull the card and nail up away from the iron filing pile. Be sure the nail point touches the index card. Some filings will be hanging underneath the index card, attracted by the magnetic field. Hold the card and nail over a second clean sheet of typing paper to catch the iron filings. Turn the electromagnet off by disconnecting one of the alligator clips on the battery. The iron filings will fall onto the paper. Lightly shake loose any filings that stick to the index card. Weigh the filings by placing the paper onto a gram weight scale. Write down the weight registering.

Repeat the experiment, but this time use 12 volts instead of 6 volts. Two 6-volt batteries can be connected together to give a total of 12 volts. Connect a jumper lead with alligator clips on each end to the positive terminal (+) of one battery and the negative terminal (–) of the second battery. Connect the two clips attached to the electromagnet to the two empty terminals on the batteries. Use fresh iron filings for the second test to avoid possible magnetized filings. Use the index card again to gather iron filings. Weigh the filings gathered. Compare the weights of the iron filings collected the first and second times.

**Something more...**

1. Is there a direct relationship between the number of turns of wire on the nail and the amount of filings it collects? For example, will twice as many turns collect twice as many iron filings?

2. What happens if you increase the voltage? Is there a direct relationship between the voltage and the amount of filings collected?

# PROJECT 1–2
# Filed Under Current
## Magnetic lines of flux and magnet's DC polarity

---

**YOU NEED**

- spool of solid conductor hookup wire
- 1.5-volt "D"-cell alkaline battery
- "D"-cell battery holder
- index card
- 3" iron nail
- powdered iron filings
- wire cutters

---

When electric current flows through a wire, a magnetic field is established surrounding the wire. This field has a pattern of lines called "flux lines," or "lines of force," by early experimenters. The pattern can be seen by sprinkling iron filings onto an index card and placing the magnetic source underneath. When the index card is tapped, the iron filings shift and align themselves with the magnetic field, and the pattern made by the lines of force becomes visible.

Does the direction (polarity) the current is flowing in the wire make a difference in the shape of the pattern? Hypothesize that the direction of current flow through a wire does not affect the pattern shape of the magnetic lines of force, that the pattern will be the same regardless of the polarity of the source voltage supplying power to the wire.

Cut a 2-foot length of solid conductor hookup wire and strip about ¼ inch of insulation off of both ends. Wrap 25 turns of wire around an iron nail about 3 inches long. Twist the positive wire coming from the battery holder to one end of the wire coiled around the nail. Connect the negative wire to the other end of the coiled wire. The reason for coiling the wire around a nail is to strengthen the magnetic field. The strengthened field makes the iron particles line up more distinctly. An "electromagnet" has just been constructed.

Lay the coiled wire-nail assembly flat on a table. Sprinkle powdered iron filings on an index card. Place the index card on top of the nail. Put a battery in the battery holder. Tap the index card several times until the iron filings form an observable pattern.

Next, without disturbing the iron filings and the coil underneath, reverse the positive and negative battery leads. You may want to use a double pole switch to reverse these leads. Observe whether or not the pattern changes and reach a conclusion about your hypothesis.

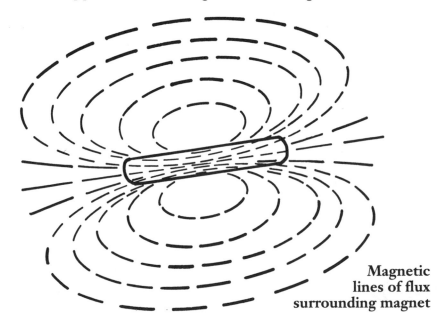

**Magnetic lines of flux surrounding magnet**

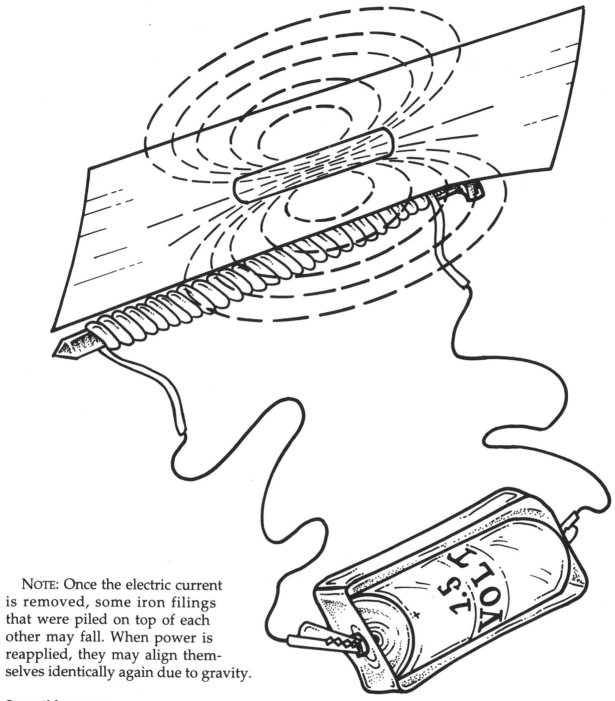

NOTE: Once the electric current is removed, some iron filings that were piled on top of each other may fall. When power is reapplied, they may align themselves identically again due to gravity.

**Something more...**

1. Why doesn't AC (alternating current) work with this experiment?

2. Does the magnetic field produce more lines of flux at the ends of the nail when the voltage to the coil of wire is increased? What about the lines along the sides of the nail?

3. Does the shape of the pattern change as the number of windings are increased, or if the nature of the wire (gauge, composition) is changed?

4. Can you diagram an "end on" field? In other words, place the magnet in a vertical position?

5. Is there a way to focus the lines of flux?

6. Put a paper clip next to a magnet. How does it affect the pattern of the lines of force?

# PROJECT 1-3
# Short Stop in a Magnetic Field
### Electrical insulators not necessarily magnetic insulators

## YOU NEED

- powdered iron filings
- wax candle
- an adult with matches
- plastic sandwich bag
- magnet
- ohmmeter
- old ashtray
- small brown-paper lunch bag
- 12-inch square of aluminum foil
- glass test tube
- iron frying pan
- water
- graphic arts materials (to make display chart)

All materials conduct electricity to some degree. Insulators are materials that do not conduct electricity well. They have a resistance so high that they block current flow. Magnetic insulators are materials which do not allow magnetic fields to pass. They shield magnetic fields and do not let magnetism affect objects on the other side.

Are materials that are insulators to electrical flow also insulators, or shields, for magnetic flow? Hypothesize that electrical insulators are not necessarily magnetic insulators.

Sprinkle some powdered iron filings into an old metal ashtray or metal lid of any wide-mouth jar. Have an adult assist in lighting a candle and dripping a few drops of the hot wax into the iron filings. The filings will stick to the wax, making several tiny balls. We will use these iron balls and a magnet to test the magnetic insulating ability of several materials.

Using graphic art supplies, make a display chart listing materials to be tested for electrical and magnetic insulation properties: aluminum, glass, iron, and water, for example.

Use an ohmmeter to check the resistance of each material and record your results on the chart. Fill in the chart with either "YES" if it is an insulator, or "NO" if it is not an insulator. If the ohmmeter needle does not move, then it is an electrical insulator. If the meter indicates the resistance is only a few ohms, then it is an excellent electrical conductor.

Next, test each material for its magnetic insulating ability. To test the plastic bag, place one of the iron-wax balls inside the bag. Place a magnet on the outside of the bag and try to move the ball inside. If the ball is able to be moved, then the bag is not a magnetic insulator. Use the same procedure to test a brown-paper bag and a glass test tube. Then fill the glass test tube with water and attempt to move the ball. To test a sheet of aluminum and an iron frying pan, hold it flat (horizontally), place an iron-wax ball on top, and try to move the ball by moving a magnet around underneath.

Study the charted results of the data. What is your conclusion of your initial hypothesis?

## Something more...
1. Test other materials such as tin, zinc, copper, rubber (use a balloon).
2. Test combinations of materials, a layer of rubber on top of a layer of aluminum, for example.
3. What affect does distance have on magnetism?
4. Test some materials that fall somewhere between being an electrical insulator and an excellent conductor.

# Rust in Pieces
### *Comparing magnetic properties of iron and rust*

When iron is exposed to damp air, a chemical process takes place, changing the iron to "iron rust" (hydrated iron oxide). The process of oxygen in the air combining with iron is called oxidation. During our procedure, we will collect a magnetic load of rusted filings and compare it with a magnetic load of regular filings. Hypothesize that one group will weigh more than the other or that they will be equal.

Cut two small paper cups in half sideways and throw away the top parts. Put several tea-

**YOU NEED**

- iron filings
- water
- magnet
- gram-weight scale
- index card
- two paper cups
- ice-cream stick

spoons of iron filings in one shortened paper cup, and several more in the second paper cup. Add water to the iron filings in one cup and let the filings soak for about an hour. Drain off the water and let it dry overnight.

In the morning, the rusty iron filings will be stuck together. Use a popsicle stick to thoroughly loosen the different particles.

Place an index card over the top of the paper cups containing the dry iron filings. Lay a magnet on top of the index card. Slowly lift the card and magnet. Iron particles will be clinging to the bottom of the card. Drop the iron particles onto a gram-weight scale. Measure the weight of the particles collected.

Repeat the procedure with the rusty iron filings. Is there a significant difference between the amount of dry iron filings and wet iron filings collected? If so, is your hypothesis correct?

**Something more...**
1. Measure the collected filings by volume.
2. Is iron rust a better or worse conductor of electricity? What effect does this have on outdoor metals that should remain good conductors of electricity (such as a lightning rod and its grounding system)?

# PROJECT 1–5
# Directional Flux
### *Inducing current using south or north magnetic poles*

---

**YOU NEED**

- large nail

- large bar magnet with north and south markings

- spool of solid conductor hookup wire

- DC milliammeter, capable of measuring down to several milliamperes

- wire strippers

---

**E**lectricity and magnetism have similar characteristics. Each, for example, has a polarity. Magnets have a north and south pole. Electricity has positive and a negative polarity. In magnetism, *like* poles repel each other and *unlike* poles attract. This is true of electricity, too. Current flows when there is a difference in electrical potential, but not when both are exactly alike. Hypothesize that there is a polarity in the current flow induced in a wire which is cutting the lines of force in a moving magnetic field.

Cut a 3-foot length of solid conductor hookup wire. Strip ½ inch of the insulation off each end. Wrap the wire in a spiral coil around a large nail, keeping the wire turns tight against each other and against the nail. Connect the two bare ends to the terminals of a DC milliammeter.

Using the north pole of the magnet, quickly stroke the magnet along the coil of wire in one direction only. Observe which direction the needle jumps on the meter (toward the positive or toward the negative). Using the south pole of the magnet, quickly stroke the magnet along the coil of wire in the same direction as you did with the north pole. Observe which direction the needle jumps on the meter. Was this direction opposite to when the north pole end of the magnet was used? Reach a conclusion about your hypothesis.

**Something more**...

Does it matter whether you stroke the magnet over the coils from left to right or from right to left?

18

# PROJECT 1–6
# Stroke of Good Luck
*Inducing magnetism in iron by stroking with permanent magnet*

Any metal object attracted by a magnet becomes magnetic itself, as long as it is in contact with a real magnet. If a strong magnet comes in contact with one end of an iron nail, the other end displays magnetic characteristics and becomes capable of attracting yet another nail, forming a chain. This effect is only temporary. When the strong magnet is removed from the first nail, the second nail falls away from the first.

An iron nail or piece of steel can be turned into a permanent magnet by stroking it with a strong magnet. The more the metal is stroked by the magnet, the stronger it becomes, but it can never become stronger than the original stroking magnet. How many strokes does it take for a soft-iron nail to show signs of becoming a magnet? Does a nail stroked 100 times have twice as much magnetic strength as one stroked only 50 times? What is the relationship between the number of times stroked and the strength of the magnetic field? Hypothesize that the strength of the induced magnet increases the more it is stroked.

Pour a pile of iron filings onto some paper. Place an iron nail in a small plastic bag. Avoiding the seam, hold the side of the bag tightly against the nail. Dip the nail, head first, into the filings, then lift it out. No iron particles should adhere to the bag, indicating that the nail has no magnetism. Remove the nail from the bag.

Using a strong magnet, stroke the nail 25 times. Be sure to use the same pole on the magnet and always stroke in the same direction. Start at the head end of the nail and stroke down. At the bottom, move the magnet away from the nail. Bring it back to the nail's top, touch the nail, and again stroke in a downward motion.

After 25 strokings, place the nail back into the bag and lower it into the iron filings. Slowly lift the bag straight up. Hold the bag low over a sheet of paper and remove the nail. Collect the

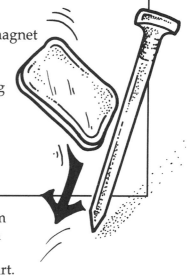

## YOU NEED

- iron nail
- strong permanent magnet
- iron filings
- plastic sandwich bag
- typing paper
- gram-weight scale
- paper and pencil

iron particles that fall from the bag, weigh them on a gram-weight scale and record the results on a chart.

Remove the filings from the scale and hold them aside. Do not put them back into the pile because they may be magnetized and interfere with the accuracy of the project.

Stroke the nail 25 more times. Again, lower the nail in the bag into the iron filings and weigh the amount collected. Record it on the chart.

Repeat the stroking and weighing procedure for 100, 500, and 1,000 strokes. Examine the data recorded on the chart. Is there a relationship between the number of strokes and the amount of iron filings attracted? Reach a conclusion about your hypothesis.

## Something more…

1. Is the relationship between the number of strokes and the amount of iron filings attracted by the nail linear or exponential?
2. How many strokes does it take before a soft-iron nail becomes observably magnetic?
3. Is there an easier way to induce magnetism? Can we simply place it in the presence of an electromagnet?

# Don't Lose Your Cool
*Magnetism loss due to heat*

## YOU NEED

- three soft-iron magnets
- strong permanent magnet
- iron filings
- typing paper
- two bowls
- plastic sandwich bag
- gram-weight scale
- cold water
- ice cubes
- boiling water

Heat removes magnetism. In this project, we will create three magnets from soft-iron nails, then attempt to note any observable changes in magnetism after being subjected to hot and cold temperatures.

Stroke a soft-iron nail 300 times with a strong permanent magnet. Always stroke it in the same direction: from the top down, moving the magnet away from the nail once at the bottom. Touch the top again and stroke down. Be sure to use the same pole of the magnet when stroking. Repeat this procedure for each of three soft-iron nails.

Place one nail into a plastic sandwich bag. Lower it into a pile of iron filings. Slowly lift the bag straight up. Hold the bag close over a sheet of typing paper. Slowly pull the nail away and collect the iron particles that fall. Weigh the iron filings on a gram weight scale. Repeat this procedure for all three nails.

The amount of filings collected by each nail must be as equal as possible. If a nail is magnetically weaker than the others, stroke it a few more times and compare the amount of filings it attracts.

Place one nail in a bowl of cold water. Add ice cubes. Have an adult boil water and pour it into another bowl. Place a nail in the boiling water. Leave the third nail at room temperature as "the control nail." Let the nails remain in the water for about one hour, until the hot water has become cooler and the cold water less cold. Repeat the procedure of using iron filing attraction to measure the strength of the nails' magnetism. Is there any observable difference in the nails' magnetic abilities? Reach a conclusion about your hypothesis.

### Something more…

1. Have an adult place a magnetized nail in an oven and heat it to 300 or 400 degrees Fahrenheit. When removed from the oven and cool, check the nail's magnetic ability.

2. Is there a relationship between the length of time the magnet is exposed to high heat and its magnetic ability? Vary time and temperature and compare magnetic attraction. Graph it on a chart.

# SECTION TWO
# STATIC ELECTRICITY

**S**liding across a fabric car seat and touching the metal door handle on a cold, dry day in winter may give you a shock. Clothes stick together when they are taken out of the dryer. These phenomena are caused by "static electricity," stationary electrical charges on objects.

Static charges can be caused by friction between two materials, such as glass and silk. Static caused by electrical disturbances in the atmosphere can be heard in an AM radio. A hissing sound in a radio, similar to escaping steam, is caused by small atmospheric charges. Strong electrical noises in a radio, sounding like loud sharp cracks, can be caused by a discharge, or arcing, of electricity in the atmosphere. Such sounds are likely to be caused by nearby electric motors and from lightning.

Static build-up occurs in materials where electrical charges do not move freely, as is true of metal conductors. The charges are called "static" because they have accumulated but are not moving. Electrons flow quickly, however, when a material with a static charge discharges, thus removing the electrical potential. Electrical potential is the difference in the charge between two materials.

# PROJECT 2–1
# Too Close for Comfort

## Static electricity on TV screens

### YOU NEED

- two cigar-shaped balloons
- 19- to 25-inch color TV set
- thread
- two chairs
- measuring stick or pole

There has long been a debate as to whether sitting too close to a TV set (or computer screen) is a health hazard. Besides light, there may be other electromagnetic fields present which may or may not be a problem for humans. This project will detect the presence of one unseen energy field that emanates from a TV set's picture tube, namely static electricity.

Blow air into a cigar-shaped balloon. Tie a thin piece of thread around its center so that it balances on the thread. Tie the other end to the center of a yardstick. Place two chairs in front of

a TV set. For best results, a larger size color TV set should be used, as larger size sets require higher picture tube voltages (typically 25,000 to 30,000 volts). Suspend the balloon so that it hovers directly in front of the TV screen. Do this by placing a rod or measuring stick across the backs of the two chairs.

First, position the chairs so that the balloon is 12 inches away from the front of the TV screen. Turn the TV on. Note any effect on the balloon, such as a movement away from or towards the TV set. Touch the balloon to a metal grounding point, such as a refrigerator or metal plumbing fixture to remove any static charge on the balloon. Repeat the procedure, poisitioning the balloon two feet from the screen, and then repeat it at a distance of three feet.

Turn the TV set off. Bring another balloon near the balloon suspended on a thread. Is there an attraction or repulsion? Does the TV set have a static electric field around it that you were able to detect by charging a balloon? Reach a conclusion about your hypothesis.

### Something more...

Measure the ability of a TV screen to charge a balloon at different distances by first exposing a balloon to a TV screen, then measuring the distance the balloon must be from a piece of thread before the thread is moved by the static field of the balloon. Start several feet away from the set, moving one foot closer each time. End by finally touching the balloon to the screen and then testing its ability to attract or repel a suspended piece of thread.

# PROJECT 2-2
# Snap, Crackle, Pop
## *Detecting approaching thunderstorms*

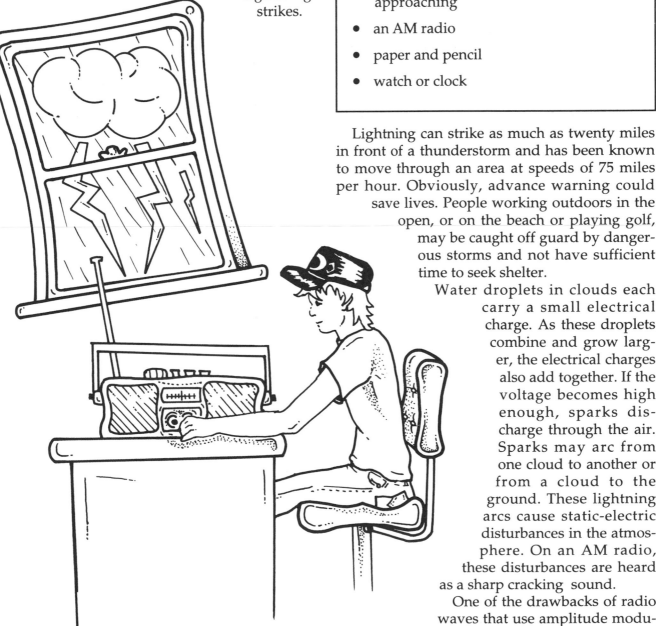

Lightning is a very dangerous natural phenomenon, taking the lives of many people each year. One popular scientific company even markets a device designed to detect approaching lightning strikes.

**YOU NEED**

- a clear day
- a day when a thunderstorm is approaching
- an AM radio
- paper and pencil
- watch or clock

Lightning can strike as much as twenty miles in front of a thunderstorm and has been known to move through an area at speeds of 75 miles per hour. Obviously, advance warning could save lives. People working outdoors in the open, or on the beach or playing golf, may be caught off guard by dangerous storms and not have sufficient time to seek shelter.

Water droplets in clouds each carry a small electrical charge. As these droplets combine and grow larger, the electrical charges also add together. If the voltage becomes high enough, sparks discharge through the air. Sparks may arc from one cloud to another or from a cloud to the ground. These lightning arcs cause static-electric disturbances in the atmosphere. On an AM radio, these disturbances are heard as a sharp cracking sound.

One of the drawbacks of radio waves that use amplitude modu-

23

lation (AM) is that the waves are easily disrupted by any kind of electrical interference in the atmosphere. This can be caused by natural electrical arcing in the atmosphere (lightning, for example) or by man-made devices like automobiles. While such arcing is annoying when we are trying to listen to the radio, it may also be of benefit if used to detect an approaching thunderstorm and serve as an early warning.

Hypothesize that you can use an AM radio to detect the approach of dangerous thunderstorms.

On a clear day, when no storms are in the forecast, tune an AM radio to a spot on the dial where no stations can be heard. For five minutes, monitor the sound on that frequency. Make a record describing any sounds you hear. If there are loud "cracks," record the number heard within the five-minute observation period.

On a day when a thunderstorm is approaching, again turn on the AM radio and tune it to a spot on the dial where no stations can be heard. (in the same position as you did for the clear-day observations). Monitor the sounds you hear for five minutes. Record the number of times

during the five minutes that loud, sharp cracks are heard. WARNING: Do not go outside when there is the threat of a thunderstorm. Perform this experiment indoors only!

Note the types of sounds and the frequency of their occurrence (the number of times they are heard in a given period of time) on the clear day and the day of an approaching electrical storm. Use your recorded data to make the comparison. Reach a conclusion about your hypothesis.

**Something more...**

1. Does the loudness and/or the number of cracks heard increase as a lightning storm gets closer to you? Use a tape recorder.

2. What sounds, if any, are heard on an AM radio when a severe storm is approaching but it does not contain any visible lightning?

3. Is TV as good an indicator of an approaching electrical storm as an AM radio? Is an FM radio as good an indicator of an approaching electrical storm as is an AM radio?

4. Can you use an AM radio to tell storm direction by knowing where the radio station is located? Does a station to the east sound clearer than one to the west?

# PROJECT 2-3
# Too "Clothes" for Comfort
### Static electricity in clothing material

The ancient Greeks were aware of the power of amber. When rubbed with certain materials, such as fur, this mineral can attract some objects that are small and light in weight. Ben Franklin first proposed the concept that static electric charges were not created nor destroyed, but were caused by a transfer of "electric fluid," as he called it, between objects. He believed in a conservation law that stated that when an amber rod is rubbed with fur, "electric fluid" is transferred from one object to the other. Although Ben did not know that this "electric fluid" as "electrons," he was correct in his principle of the conservation of charges. The charges were not created or destroyed, only moved.

With clothes dryers in many modern homes today, you may have observed some clothing sticking together as you remove them at the end of the drying cycle. Could this static electricity be caused by the hot, dry, tumbling environment of the dryer? Hypothesize which types of materials are more apt to become statically charged in the clothes dryer.

Gather an assortment of clothing for washing and drying. Include as many different types of fabric materials as you can. Along with cotton, wool, and silk, add such synthetic-fiber fabrics as polyester, nylon, and acrylic. Wash them and then dry them in a dryer. (NOTE: Some laundry products are designed to remove static electricity.) Record which types of materials appear to have a static charge. Note which materials attract and repel other materials. Remember that like charges repel and unlike charges attract. Reach a conclusion about your hypothesis.

### Something more...
1. Since unlike charges attract and like repel, can you determine which types of material are similarly charged and which have opposite charges?
2. Are there any natural substances that you

---

### YOU NEED

- an assortment of clothes made from different fabrics (Include nylon, acrylic, polyester, silk, cotton, and wool items)
- use of a laundry dryer

---

could stick in the dryer to reduce static cling, other than the commercially available "static-free" products? How effective are these various in-dryer products that are designed to make clothing static-free? Are they effective in reducing static electricity in all types of fabrics?

3. It has been said that dish towels don't absorb moisture as well once they are put through a drying cycle in which a commercial static-free product was used. Can you confirm this?

4. When two pieces of clothing stick together after coming out of the dryer, do the static charges soon equalize so that the two articles of clothing lose their attraction?

5. Do clothes in a dryer develop the static charge as moisture leaves, or would dry clothes put through a cycle develop a charge too?

# PROJECT 2–4
# Grounded Again
### *Removing static charges in clothing by grounding*

---

**YOU NEED**

- several articles of clothing made from either nylon, polyester, rayon, or acrylic
- use of a laundry dryer
- metal coat hanger
- alligator clips with screw terminals
- roll of solid core hookup wire
- 2- or 3-foot-long piece of metal pipe
- hammer
- wire cutters
- access to patch of soft ground dirt in the ground

---

The ground (Earth's surface) is considered to be a point of zero volts. When we "ground" a piece of electrical equipment, it means that an electrical connection is made from the piece of equipment to the Earth. This is often done for safety reasons, for example, connecting a washing machine's metal cabinet to a ground point. Then, if a "hot" electric wire in the washer, one carrying a high voltage, should come in contact with the cabinet, the person using the washer would not receive an electric shock or be electrocuted. The high voltage drains off to the Earth grounding.

Each electrical outlet box in your home is supposed to be connected to a wire that goes to a grounding point. Lightning rods on houses are grounded by means of thick wire from the rod on the roof to a metal pipe or copper rod driven several feet into the ground.

When clothes come out of the laundry dryer, some fabrics have an annoying static buildup in them. Will touching such fabrics with a conductor connected to an Earth ground drain the static electricity out of the clothing? Wouldn't it be great if this static could be discharged simply and quickly without the use of commercial products or chemical sprays? Hypothesize whether or not an Earth ground can be used to remove static electricity from clothing.

Wet several articles of clothing made from such synthetic materials as nylon, polyester, rayon, and acrylic, materials which commonly display static cling when removed from a dryer. Place the clothes in the dryer and turn it on.

Using a hammer, drive a 2- to 3-foot-long metal pipe into the ground, leaving only a few inches extending above ground. Cut a 4- or 5-foot length of solid-core hookup wire. Strip about ¾ of an inch of insulation from both ends of the wire. Install an alligator clip on each end. Connect one alligator clip to the metal pipe and the other to a metal coat hanger.

Remove the static-charged clothing from the dryer. Slowly wipe the grounded metal coat hanger over each clothing item. Did the hanger remove any of the static? Did garments that were clinging together separate? Reach a conclusion about your hypothesis.

**Something more...**
1. Can you quantify the amount of static built up in a garment by measuring the distance of an arc? (This might best be seen, and enjoyed, in a darkened room or outside at night.)
2. How might the pipe-ground connection be improved? Would soaking the ground around the pipe with water make it ground better?
3. Are some fabric materials harder to discharge than others?

# PROJECT 2-5
# Tea for Two
## *Electrostatic generation (Van de Graaff generator)*

A common device found in many school science rooms is the "electrostatic generator," also known as a "Van de Graaff generator." This device is used to build up a high electrical charge ranging from about 200,000 volts, from a relatively small unit, to much higher voltages for bigger ones. A motor drives a belt made of gum or other material inside a tube. A smooth round dome is placed on the top. The dome is made large so that the electricity does not leak off easily. As the belt moves at high speed, a grounding wire drains negative charges off the belt at the bottom of the device, thus leaving a positive charge on the top dome. The Van de Graaff generator is a good continuous source of static electricity.

Electrostatic fields are used in many applications, such as the removal of particles from the air (dust, pollen, or other particulate matter in industrial chimneys) and to direct the flow of sprayed paint.

Hypothesize that, since like charges repel, two strings taped to the top of a Van de Graaff

dome will repel each other because they will be charged to the same potential.

The Van de Graaff generator is capable of giving anyone who touches it an unpleasant shock, but because there is little current it is not harmful. Avoid touching the generator while it is on. If you must touch it, use a long stick, such as a broom handle or a wooden back-scratcher.

Remove the string and tags from two tea bags. Place the tags next to each other and tape them to the top of the Van de Graaff generator's dome. Turn the generator on. Do the strings appear to stand up straight? Are they standing perpendicular to the dome? Do they lean toward each other, or away from each other? Reach a conclusion about your hypothesis based on your observation.

### Something more...

1. Wet the strings. What effect, if any, does this have?
2. What happens when you bring another object into the field, such as a wooden back-scratcher?
3. Is the three-dimensional static field surrounding the dome symmetrical? Is it higher than it is wide?
4. Light a candle and place it next to the generator. Why does the flame bend as though a steady wind was blowing from the dome?
5. Blow soap bubbles across the top of the generator. Hypothesize whether the bubbles will be attracted to the dome or repelled as they approach it.

# PROJECT 2-6
# Static North Pole
### Magnetic properties of static electrons

---

### YOU NEED

- magnetic compass
- Van de Graaff generator
- hardbound book

---

It is common knowledge that electric current flowing in a wire exhibits magnetic properties. A magnetic field can be detected surrounding the wire. This law of nature makes electric motors and generators possible. But how about static electricity? Does it have magnetic properties, too? Hypothesize that static electricity also has magnetic properties.

To prove our hypothesis, we must establish a static field and detect the presence of a magnetic field. A simple compass will serve as our tool for detecting a magnetic field. Many school science classrooms have Van de Graaff generators, a device which creates static electricity.

Place a compass on a hardbound book. You will get an unpleasant shock if your hand comes too close to the generator. The book will be an insulator, allowing you to move the compass near the generator.

Turn the compass so that the north-pointing half of the needle is directly over the north marking on the compass's face. While the generator is still off, bring the book and compass close to the top dome of the generator. The needle should not move. The dome is made of aluminum and should not affect the magnet.

Turn the generator on. Slowly move the book and compass close to the top dome. Is the needle deflected from its normal north-seeking direction? If so, then your hypothesis was correct.

### Something more...

Hypothesize whether electrons are polarized to magnetic north or magnetic south. This test can be done by evaluating the results of the experiment above. The top dome of a Van de Graaff generator is supposed to have a positive charge because a grounding wire gap at the base of the generator drains electrons from the rotating belt. The dome has a lack of electrons, and thus has a positive charge. Recall that like poles repel and unlike poles attract. Therefore, the north-seeking half of a compass needle is the south pole part of the needle. If the compass needle points to the dome, then the south pole is being attracted to a positive charge. Therefore, electrons would be attracted to magnetic north, and since unlike polarities attract, then electrons would be assigned south pole properties.

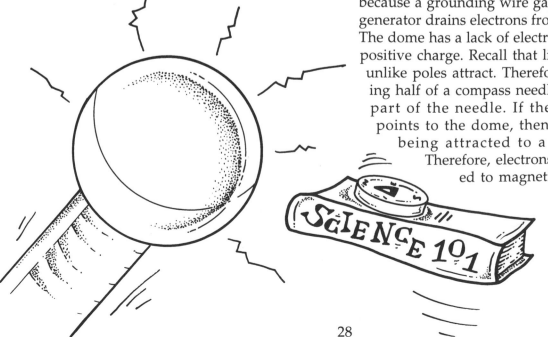

28

# SECTION THREE
# CURRENT FLOW

Electricity is described using two terms; potential and current. Current flow is the movement of electrons (negatively charged particles) through a medium. The medium is called a "conductor" because it conducts, or carries, the current along a path. Ordinary wire, with which you are familiar, is a conductor of electricity. We often use the phrase "the flow of electricity" to mean the flow of current, both terms being identical. Current is measured in "amperes." Small current flows are measured in "milliamperes," a milliampere being 1/1000th of an ampere. Amperes and milliamperes are commonly shortened to simply "amps" and "milliamps" when writing or talking about electricity. These are measures of the strength of the electricity's flow.

"Electrical potential" is a measure of stored electrons with the possibility of doing work. There may be no work being done, but there is the "potential" to do it. Electrical potential is measured in "volts." A new flashlight battery sitting on the table has a difference in electrical potential of 1.5 volts across its positive and negative terminals.

Current flows when there is a difference in "electrical potential" between two points and a conductor is connected between them. The conductor provides a path for electrons to move. One point has too many electrons gathered and another has too few. Similar to water seeking its own level, electrons move along from an area where there is a lot of electrons to where there are a lesser amount.

# PROJECT 3–1
# Volts and Amps
### *Understanding the concepts of voltage and amperage*

## YOU NEED

- 1.5-volt "D" flashlight battery
- "D"-cell battery holder
- incandescent "flashlight" bulb
- bulb socket/holder
- insulated jumper leads, with alligator clips on each end
- voltmeter
- milliammeter
- on/off switch

**B**efore a person can do more involved electrical projects, it is important that there be a good understanding of the two terms "potential" and "current flow." This project, aimed at lower grade levels, is a demonstration which will help fellow classmates understand these two terms.

Connect a battery to a switch and a small lamp (flashlight bulb). Voltmeters measure the difference in potential across two terminals. Place the positive probe of the voltmeter on the top of the battery and the negative on the bottom. The meter is said to be "in parallel" with the battery. It will read about 1.5 volts, regardless of whether the lamp is lit or not. The milliammeter must be placed "in series" (in the path) with the current flow. It measures how much current is flowing through the circuit. In this setup, the lamp is drawing the

power to light it. That power is measured by the milliammeter.

Throw the switch to "open" the path. The voltmeter will still show potential but the milliammeter will drop to zero.

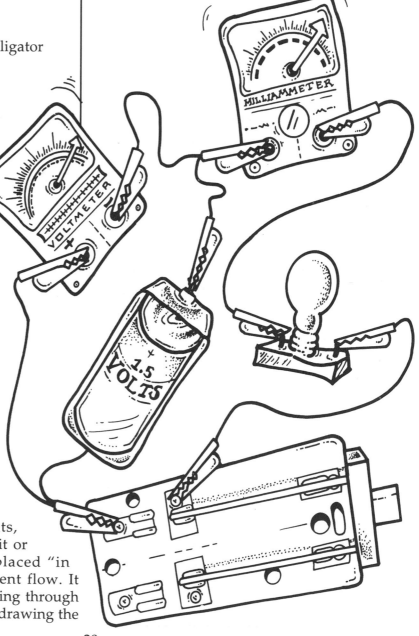

# Wet-Cell Battery
## Testing electrolytes to make wet-cell batteries

Electricity can be produced by chemical action. Take a piece of paper and wet it. Make a sandwich by placing a nickel on the bottom, the wet piece of paper in the middle, and a penny on top. Here we have two dissimilar metals (called "electrodes") and an "electrolyte" substance between them. An electrolyte is a liquid solution which conducts electricity and, when used with certain metals, can generate electricity. Electrolytes are typically made up of weak sulfuric acid or dissolved salt. Place a milliammeter probe on the penny and the other probe on the nickel. The meter's needle will move, indicating a current flow.

Hypothesize that while pure water is best for drinking, it is among the worst electrolytes.

Place a piece of zinc and a piece of carbon in a beaker or wide-mouth jar. These will be the electrodes. You may wish to use a little masking tape to secure the electrodes to the side of the beaker. Attach jumper wires with alligator clips from one electrode to a terminal on a milliammeter and another from the meter to the other electrode. The zinc electrode will have a negative charge and the carbon will be positive. The flow of electricity is in one direction. Will it matter which terminal on the meter (the positive "+" or the negative "–") is connected to the zinc or carbon electrodes?

Fill the beaker with 400 millilitres of lemon juice. Read and record the milliammeter reading. This is 100% lemon juice.

Pour out 200 millilitres of the juice, and replace it with 200 millilitres of distilled water. Stir. This makes a 50% lemon-juice solution. Record the meter reading.

Rinse the beaker and fill it with 400 millilitres of distilled water. Record the meter reading.

Can this device be used as a water tester? Is pH (the measure of alkalinity/acidity) a factor in determining a good electrolyte? Try raising

### YOU NEED

- a piece of zinc
- a piece of carbon
- 500 milliliters of citric acid (lemon juice)
- a glass beaker or wide-mouth jar
- insulated jumper leads with alligator clips on each end
- a milliammeter
- distilled water
- spring water

the pH of the water with potash or some other alkaline solution. Is it true that the purest drinking water makes the poorest battery? Reach a conclusion about your hypothesis.

### Something more...

1. Try many different electrolyte solutions: club soda, spring water, lake or sea water, apple juice, orange juice, vegetable oil. Crystals of these substances can also be dissolved in water to make electrolyte: potassium chloride, sodium chloride (table salt), as well as bromium chloride (sea salt).

2. Can you construct several wet-cell batteries and place them in series (similar to the two batteries in flashlights) to increase the overall voltage potential?

3. Hypothesize that larger electrodes produce more current than smaller ones because there is more electrode surface area in contact with the electrolyte.

# PROJECT 3-3
# After the Warm-Up
### The effect of temperature on battery life

---

## YOU NEED

- three 1.5-volt "D" flashlight batteries
- three "D"-cell battery holders
- three incandescent "flashlight" bulbs
- three bulb socket and holder
- hook up wire or jumper leads with alligator clips on each end
- use of a baseboard heater, hot air duct, or sunny window
- use of a freezer
- three thermometers
- log book and pencil
- clock

---

When the icy cold winter winds howl, you will undoubtedly hear someone complain that they couldn't get their car started that morning. Often the cause is a bad car battery, affected by the cold. Does temperature affect the dry cell batteries found in flashlights, toys, and portable radios? Does operating them in the cold enable them to generate energy longer than at room temperature? If so, does heat have the opposite effect?

Hypothesize that heat extends the hours of useful energy produced by a flashlight battery and a decrease in temperature causes a decrease in battery life (or hypothesize that the exact opposite happens).

Connect a light bulb (lamp) to a battery. The hookup wire between them should be about a foot long because we are going to place one of the batteries inside a refrigerator's freezer but we want the bulb to stay outside at room tem-

perature. Construct two other similar setups to give a total of three battery/lamp devices.

Place one device on a table at room temperature. Place the other in an area of higher temperature, such as a sunny window or on a hot-air duct. You can put it near an electric baseboard heat unit or radiator, but not too close. (For safety, have an adult help you place this battery device in an appropriately warm area.) The third battery will be placed inside a freezer. Let the hookup wires from the battery to the bulb extend out the freezer door and connect to the bulb outside, which will remain at room temperature. It may not make a difference, but generally the electrical resistance of wire is decreased as temperature decreases. The short piece of wire and small range of temperatures we are dealing with should not make enough difference in wire resistance to affect the results of your project. But the filament of a light bulb is a special kind of wire so we will keep the bulb out of the cold (and the other bulb out of the direct heat) to give our experiment the most accurate result.

When you construct the three battery/lamp devices, make one of the wires easy to connect and disconnect. If you use a battery holder, you can easily disconnect the wire, as you can if you use alligator clips.

Place the three devices at their locations (room-temperature table, freezer, and warm area). Connect the batteries to the lamps. In a log book, record the time on the clock that the experiment begins. Every half hour, check each lamp to see if it is still burning. Since this project will take several days, disconnect all of the batteries from the lamps (only one wire needs to be disconnected to open the circuit) when you want to stop to go to school or to bed because it is necessary to constantly observe the performance of each set. If you forget, and one battery stops just after you go to bed, and another stops just before you get up, you would not know which one lasted longer. Note the time you disconnect the batteries in your log book and calculate the number of hours each has been lit. The next time you will be home for a while, continue your experiment by hooking all of the batteries back up and start your half-hour monitoring again.

Record which battery gives out first, second, and third. Compare this to the statement you made as a hypothesis to determine if that statement is correct or incorrect.

### Something more…

1. Does temperature affect the shelf life of a battery? Is it better to store new batteries in the freezer instead of a cabinet drawer?

2. Measure the lumens (a measure of light brightness) of each lamp using a light meter. Do they decrease and slowly die out or just die out all at once? Will peak output be maintained longer by one of the three batteries?

3. Alter the gauge of the wire from very thin to very thick. Hypothesize an effect.

4. In the above project, the lamps were kept out of the warm and cold areas. Repeat the project with the lamps *in* the warm and cold areas as well as the batteries.

5. Increase your sample size (use six batteries instead of three).

6. Does intermittent use of a battery extend its life? Maybe a battery recovers a little if it is given a rest period every hour, which could mean a longer total useful life.

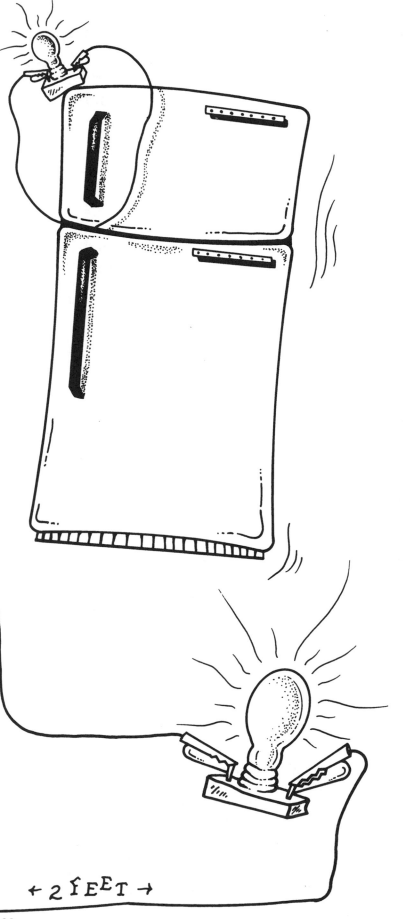

# PROJECT 3-4
# Keeps On Keeping On
*Comparing popular brands of batteries*

## YOU NEED

- three 1.5-volt "D" cells from different manufacturers (All battery brands used in test—Duracell, Eveready, Radio Shack, Mallory, Ray-O-Vac, or other—must be made of same material, such as alkaline.)
- "D"-cell battery holders
- incandescent flashlight bulbs (lamps), one for each battery
- bulb socket/holders, one for each battery
- hook-up wire or insulated jumper leads with alligator clips on each end
- log book and pencil
- clock

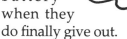

All battery commercials and ads boast that their brand is the best. But they can't all be the "best." One might have the best price. Another might have the best shelf life. Yet another may be best at providing power for a longer amount of time.

Do you believe that some popular name-brand batteries really do run longer than other brands? Hypothesize which batteries you believe produce useful power for the longest time. This project assumes that all batteries are equally new and fresh.

Connect a lamp to a battery. Do this for each battery you have. Turn all of the lamps on at the same time. After many hours (or days) of operation, watch for dimming. When the lamps begin to dim, start checking every half hour. Since this project will require several days, disconnect each lamp from its battery during periods when you cannot monitor them (such as while you sleep or are at school). In a log book, record the time you start and stop the project each day so that you can work out the total operating time for each battery when they do finally give out.

Study your log book records and determine if the most-popular brand-name batteries really do last the longest.

## Something more...

1. Are the results the same if you use standard-formula batteries rather than alkaline ones (or vice versa)?

2. Are the results the same for batteries of other sizes: "C" cells, "AA," "AAA," "N," 9-volt?

3. Compare cost effectiveness. Do the batteries that last the longest also have the best price? Determine the cost per hour to light your lamp for each battery manufacturer.

4. Is the discrepancy in usage greater within the brand or from one brand to another? In other words, are there greater differences among one manufacturer's batteries compared to differences among many manufacturers?

# PROJECT 3–5
# Live-Wire Wood
### Insulators become conductors under certain conditions

Some objects that are normally insulators can become conductors of electricity under certain circumstances. If someone were trapped by a downed "live wire" from a telephone pole, a person might think a tree branch or broom handle could be used to safely push the electric wire away, since wood is an insulator. But wood soaked from being in a storm or lying in water will conduct electricity.

Hypothesize that objects which normally insulate can become conductors when wet and, therefore, may not be safe to use in handling electricity.

Using an ohmmeter, check the resistance of objects that are normally insulators. Use as many objects as you can think of, such as a toothpick or pencil (representing wood), a wax candle, and a piece of rubber (rubber glove used to wash dishes). Record their resistance in a log book or chart. They should all have such a high resistance that the meter's needle will read infinity. This means they will have a very low conductivity rate (poor conductors). The

### YOU NEED

- toothpick
- wax candle
- piece of rubber
- ohmmeter
- jumper leads with alligator clips on each end
- log book, pencil
- tap water

higher the resistance, the lower the conductivity.

Soak the objects in water for a few minutes. Again, read the resistance of each object and record the results. Try moving the meter probe points close together on the objects and at opposite ends. Note if there is any change in resistance with distance across the surface of the object. Study your results and come to a conclusion about your hypothesis.

### Something more...
1. Repeat the experiment adding salt to the water. What would happen if these objects had become wet from salty sea water?
2. Weigh a small piece of wood. Soak it for an hour in tap water (place a brick or other weight on it to keep it under water). Weigh it again. Let the wood thoroughly dry out for several days, perhaps placing it in a sunny window. Again, weigh the wood when it is dry. Repeat the procedure with another piece of wood using salt water. Does the wood absorb more salt water then fresh water?
3. Try balsa wood, pine, oak, and an assortment of wood densities. Does the density of the wood affect the conductivity when wet?

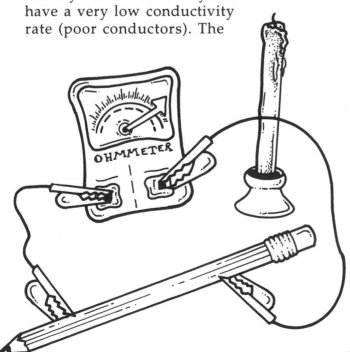

# PROJECT 3-6
# The Match Box
### Series circuit becomes teaching aid for other subjects

### YOU NEED

- 20 small bolts about 1 inch long with 40 matching nuts
- piece of 1/8-inch plywood about 2 feet by 3 feet
- spool of solid conductor hookup wire
- 9-volt alkaline battery
- 9-volt battery clip connector
- 9-volt battery holder
- several small wood screws
- piezo buzzer
- black electrical tape
- art paints
- map of the United States for reference
- hand drill
- wire cutters

A "series circuit" is where the electrical components are connected together forming one path. The path of electricity can be visualized as starting at a battery (the source of the electricity) and then following along a path made of wire to the next component, perhaps a switch, and then to the next, a light bulb for instance, and finally back to the other battery terminal. This can also be thought of in terms of water travelling in a path through pipes.

Simply touching two wires together to complete a series circuit can make a bulb illuminate or a buzzer sound when a battery is in the circuit. In this project, a series circuit is used as an educational tool for drill, practice, and testing a student on identification of states in the United States. (Countries on a continent could also be used.) This is a classic demonstration for young students, but older students should pay attention to the **Something More** ideas for more serious science spin-offs of this project.

Hypothesize that you can construct a project that will not only demonstrate the concept of a series circuit but also provide an educational tool for learning and testing such subjects as geography.

On a thin piece of plywood about 2 by 3 feet, paint an outline of the United States or a continent and an outline of each state or country. Leave room on one side of the board for a vertical list of target names. Leave an area about 4 inches square at the bottom of the display board at the opposite side. A battery and piezo buzzer will be mounted here for testing.

Select 10 states or countries that you think most students in your school would have trouble identifying or locating and could use this project to help them learn. Use a hand drill to make a hole in each of the 10 areas you have selected. The size hole will depend on the size of the bolt you are using. (If only a power drill is available to make the holes, have an adult assist you.)

To one side on the front of the board, use markers or paint to list the 10 names you have selected. You may opt to print the names on a typewriter or a computer printer and paste them onto the board to enhance your display's appearance. List the names in alphabetical order. Drill 10 holes, one next to each name.

Push 20 bolts through the holes in the board, with the bolt head on the painted side of the board. Use 20 nuts to secure them to the board.

Using hookup wire, run a wire along the back side of the display board from the bolt in each site to its corresponding name in the name listing. There will be a total of 20 pieces of wire, one

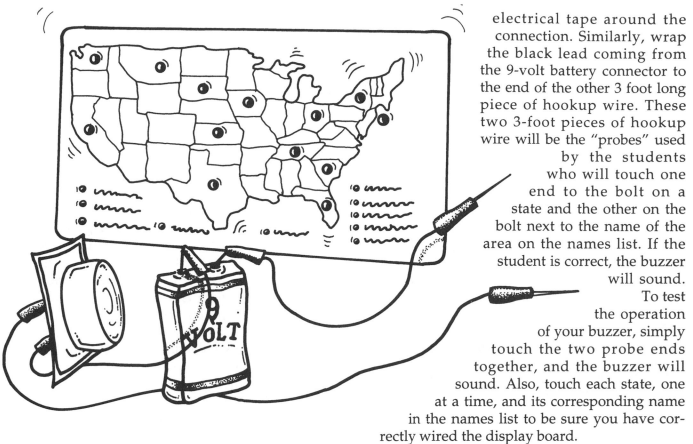

electrical tape around the connection. Similarly, wrap the black lead coming from the 9-volt battery connector to the end of the other 3 foot long piece of hookup wire. These two 3-foot pieces of hookup wire will be the "probes" used by the students who will touch one end to the bolt on a state and the other on the bolt next to the name of the area on the names list. If the student is correct, the buzzer will sound.

To test the operation of your buzzer, simply touch the two probe ends together, and the buzzer will sound. Also, touch each state, one at a time, and its corresponding name in the names list to be sure you have correctly wired the display board.

Have several classmates test their knowledge of geography using your display board. Ask teachers in your school for their opinions about the educational value of your project, and reach a conclusion about your hypothesis.

for each location, some will be longer than others. Using wire cutters or a wire stripper, remove about a half-inch of insulation off of each end of the wire. Each wire end can be fastened to a bolt by looping the bare wire once around the bolt on the back of the display board and screwing another nut onto the bolt. With the wire wrapped around the bolt between the two nuts, the two nuts will hold the wire securely.

Next, mount a 9-volt battery at the bottom side on the front of the display board. A battery holder and two wood screws will hold the battery securely. Push a 9-volt battery clip connector, with two leads extending from it, onto the top of the battery. Using two wood screws, mount a piezo buzzer next to the battery. Twist the exposed bare wire ends of the two red wires together, namely the red positive wire coming from the battery connector to the red positive lead (wire) on the piezo buzzer. Wrap a small piece of black electrical tape around the connection.

Cut two pieces of hookup wire, each about 3 feet in length, and strip about a half inch of insulation off of each end. Wrap the black lead coming from the piezo buzzer to the end of one of the pieces of hookup wire. Wrap a piece of black

### Something more...

1. Make the board much bigger to accommodate all 50 states or a world map. You will then need many more small bolts and matching nuts.
2. This project can be used to help teach any subject where a match between two items must be made; countries or states to their capital cities, scientists to their discoveries, chemical names to their chemical symbols (iron/FE, for example), plants or animals to their species, rocks and minerals to their significant characteristics.
3. Using the board constructed in the project above, hypothesize that people who watch the news or weather faithfully every night will be more aware of geography than those who seldom watch, and will score higher when given your test.
4. Construct a similar project so that if a person touches the correct two points to make a match, a light illuminates, but if they connect the wrong two a piezo buzzer sounds.

# PROJECT 3-7
# Current Event
### *Fuses protect from excess current flow*

## YOU NEED

- two 1.5-volt "C"-cell alkaline batteries
- 1-amp fast-acting fuse
- dual "C"-cell battery holder
- single-pole single-throw toggle switch
- LED (Light Emitting Diode)
- insulated jumper leads with alligator clips on each end

A "short" in an electrical circuit can cause an excess amount of current to be drawn from the power source. A "short" circuit is when the resistance of a circuit falls from its normal value to a very low value. This may occur when an electrical component fails. Appliances such as electric heaters, microwave ovens, and hair dryers, normally draw a considerable amount of current. A short circuit in such an appliance could cause an excess amount of current to be drawn from the electrical outlet. This condition, if allowed to continue, could cause the AC power cord on the appliance (and even the wiring in the walls of the house) to heat up. Such heat places surrounding materials in danger of being set on fire.

To prevent an appliance or electronic device from drawing too much current, a safety component, the "fuse," is often used. Fuses, which are also used in electrical panels to protect the wiring in homes, can protect the power supply from having too much current "pulled" from it. It can also protect other components inside the electrical device from damage.

Fuses in electronic devices come in many shapes and forms. The most common is a small glass tube, looking like a one-inch length of soda straw, with a metal cap on each end. Inside the tube a strand of wire, made of an alloy which melts at a low temperature, completes the circuit and allows current to pass. If too much current flows, the wire begins to heat up. When the wire reaches a certain temperature it melts, and the electrical path is broken. This open circuit safely stops the current flow and protects the power source as well as other components inside the appliance or electronic device. Fuses are rated by the number of amperes they will allow to flow through them before they open and stop the flow.

Hypothesize that a fuse can safely protect a circuit from overloading the power source to which it is connected.

Construct the circuit shown in the illustration. Be sure the toggle switch is in the open (turned off) position before the batteries are connected to the circuit. Upon connecting the batteries, the LED will light. Under this normal condition, the LED requires less than one ampere of current to illuminate it. When closed (turned to the on position), the toggle switch placed across the LED will "short out" the LED, that is, it will cause the LED to be bypassed, and current will no longer flow through the LED. This is due to the fact that electricity takes the path of least resistance. The switch provides a path of almost no resistance. Closing the switch removes the "load" (the power-absorbing device), which was the LED, and the circuit becomes a "dead short." Effectively, there is now no resistance in the circuit to impede the flow of current. Therefore, the circuit will try to draw as much current as the batteries can supply, and alkaline batteries are capable of delivering a heavy supply, more than one ampere. As the current rises above one ampere, the fast-acting fuse will "open," breaking the flow of current, and the batteries will be protected from being drained.

Perform the experiment and record your observations. Reach a conclusion about your hypothesis for the data gathered.

**Something more…**

1. What happens if the 1-amp fuse is replaced with a 2-amp fuse? A 3-amp fuse? A 10-amp fuse?

2. Place a 500 milliammeter in series with the fuse and batteries. How many milliamperes does the circuit draw normally when the LED is lit? When the switch that shorts across the LED is closed, does the milliammeter register an increase in current flow the instant before the fuse blows?

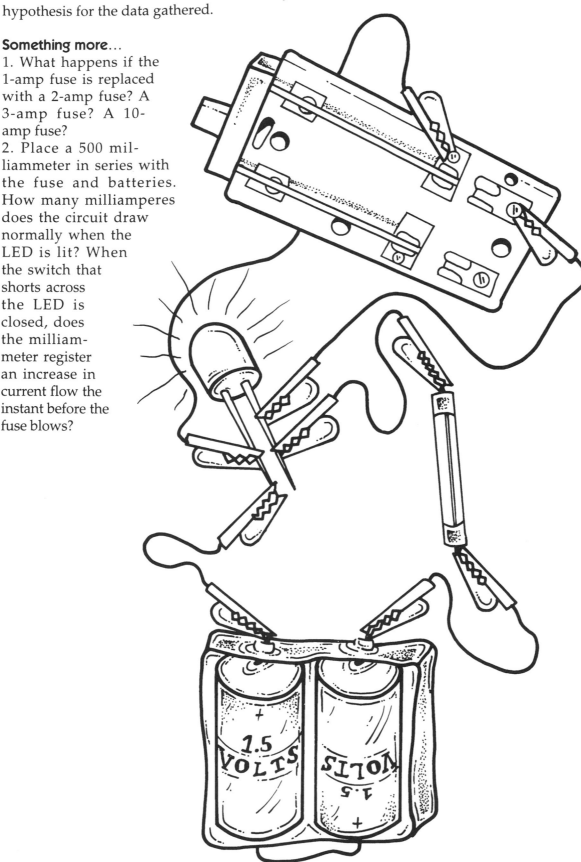

# SECTION FOUR
# ELECTROMECHANICAL DEVICES, MOTORS

We are living in a machine age. Motors are everywhere. In the home, they can be found in swimming pool pumps, refrigerators, washing machines, clothes dryers, hair dryers, VCRs, electric lawn mowers, fans, and water pumps. All motors operate due to electromagnetic principals. Basically, a motor converts electricity into motion; a conversion of energy. A generator does the opposite. Motion produces electricity.

Magnetism is a force that acts between certain materials (or devices). When electricity passes through a wire, a magnetic field is created surrounding the wire. A magnetic field in motion will produce a voltage potential in a wire. We can cause some metal materials, such as soft iron, to become permanently magnetized by placing them in an electromagnetic field.

# PROJECT 4-1
# All Coiled Up

*Effect on current flow by magnet moving past a coil*

## YOU NEED

- cardboard toilet-tissue or towel roll

- spool of solid core hookup wire

- DC milliammeter capable of measuring to several milliamperes

- a rectangular magnet just small enough to fit through a cardboard toilet tissue roll regardless of how it is positioned

- wire cutters or wire strippers

- insulated jumper leads with alligator clips on each end

Scientists had long suspected that there was a connection between magnetism and electricity, because both have similar properties. Polarity is one of these properties. There are north and south poles in magnetism just as there are positive and negative polarity in electricity. Also, in both cases, opposites attract.

In 1820 Hans Christian Orsted, a Danish physicist, discovered that a compass needle moved when brought into the presence of a wire in which current was flowing. The reverse also happens. Electrical energy is induced in a wire when that wire moves through a magnetic field.

Is it enough to simply move a magnet near a wire or coil of wire to induce an electrical flow? Hypothesize that doing that, indeed, is not enough. The polarity of the magnet in relation to the wire's position is an important factor, too.

Using a cardboard toilet-tissue roll or a section cut from a paper-towel roll, wrap 50 turns of solid-core hookup wire around the roll in a spiral pattern. Strip about an inch of insulation off both ends of the hookup wire. Connect each end of the coil of wire to the terminals on a DC milliammeter using insulated jumper cables with alligator clips.

Hold the magnet in a horizontal position and drop it down through the cardboard roll. Record the number of milliamperes that registered on the meter. Rotate the magnet 90 degrees (hold it vertically) and drop it again. Record the number of milliamperes that registered on the meter. When the meter's needle moves, it will only jump for the briefest second. Record the highest number the needle reaches during its momentary jump.

Compare the readings of the current from both drops through the tube. If a higher reading occurred when the magnet was dropped in one particular position, then your hypothesis was correct.

## Something more...

1. What results are obtained when the number of turns of wire is increased to 100 turns?
2. What results are obtained when thicker diameter wire is used to make the coil?

# PROJECT 4–2
# Marvelous Magnetic Measurer
## Detecting magnetism

How can you detect if an object has magnetism? A strong magnet can, of course, attract soft-iron nails, paper clips, and may even adhere itself to a refrigerator. But there may be times when you want to detect even weak magnetism in an object. Project #1–6, Stroke of Good Luck, for example, demonstrates an attempt to magnetize an iron nail. To do this, the nail is stroked many times with a permanent magnet. How many times must it be stroked before it displays *any* observable magnetic effect? To answer this, a device sensitive to even a small amount of magnetism is needed. Hypothesize that you can create a testing device that can detect small amounts of magnetism in objects.

Using several wood screws or nails and some short lengths of wood, construct a small stand about one to two feet high. Take a desk stapler and crimp a staple onto the end of a piece of thread. Unroll a length of thread about 2 feet long and cut it off the spool. Tie it around the top of the wood stand so that staple hovers suspended about an inch from the stand bottom.

The mass of the staple is very small, and since it is suspended on a thread, it doesn't require much force to make it move laterally (sideways in any direction). This makes it very sensitive.

Slowly bring a soft-iron nail near the staple. The staple should not move in reaction to the nail. If it does, then the nail has already

been magnetized and you must replace it with another nail.

Once you have established that the nail has no observable effect on the staple, it is ready to be stroked with a permanent magnet. Stroke the nail once, starting at the top of the nail and stroking downward. Bring the nail close to the staple and see if it has taken on any signs of magnetism. Stroke it a second time, and repeat the test. Be sure to use the same pole on the magnet each time you stroke. Continue stroking and testing until the staple is moved by the nearness of the nail. How many strokes does it take before the proximity of the nail has an observable effect on the staple? Reach a conclusion about your hypothesis.

### Something more...

Quantify the amount of magnetism in your test samples by measuring the distance the staple is deviated from plumb (moved from the center hanging point).

# PROJECT 4–3
# Currently Set Up
## *Magnetic induction*

---

**YOU NEED**

- spool of solid-conductor hookup wire
- sensitive DC milliammeter
- soft-iron nail
- 6- or 12-volt lantern battery
- insulated jumper leads with alligator clips on both ends
- single-pole, single-throw, normally open, momentary contact switch
- wire cutters

---

When electric current flows through a wire, a magnetic field is established around the outside of the wire. If another wire is brought into the magnetic field, electric current is induced into the second wire by magnetic induction.

In this project, a coil of wire is wound around a nail and a second wire is coiled over it. When a battery is connected to the first coiled wire, known as the "primary" winding, a sensitive DC milliammeter connected across the "secondary" winding will register a jump; ever so slightly and for only an instant but it will be detectable.

Although only a slight momentary movement of the meter's needle occurs, the discovery of this effect ranks among the most important of the electrical age. The concept of electromagnetic induction enabled the invention of coils, transformers, motors, and generators.

Although two scientists, Michael Faraday and Joseph Henry, working independently, discovered this electromagnetism effect at the same time, it is Faraday who receives the credit as he was the first to publish his findings.

The year was 1831, when Faraday wound two unconnected coils of wire onto a donut-like ring made of iron. Iron was used to intensify the magnetic field. When a battery was connected across the first coil, the needle of his galvanometer (a gauge that measures current flow), which was connected across the second winding, reacted with a momentarily jump. The amount of current induced into the secondary winding was very tiny, but he had proven that current could be induced in a wire through the use of magnetism.

Hypothesize that you can duplicate Faraday's results, inducing the momentary current flow in a wire and thereby prove, again, the magnetic-induction process.

Construct the circuit as shown in the illustration on the opposite page. Wrap 50 turns of wire around a soft-iron nail and strip about ¼ inch of the insulation off each end of the wire. Wrap a second coil of wire around the nail (also 50 turns) over the top of the first coil. Strip about ¼ inch of insulation from each end of this wire, also.

Connect the ends of the second (top) wire coil to a sensitive DC milliammeter. Connect a momentary contact switch in series with a battery and the first (under) coil. Push the button. What happens to the needle on the gauge at the moment the button is pushed? What does the meter's needle do when you continue to hold the button down? What happens when you release the button? Reach a conclusion about your hypothesis.

## Something more…

1. What can be done to increase the deflection of the meter's needle, to make the effect more pronounced? Does increasing the number of turns of the wire help? What happens if the number of coil turns of the primary winding is more or less

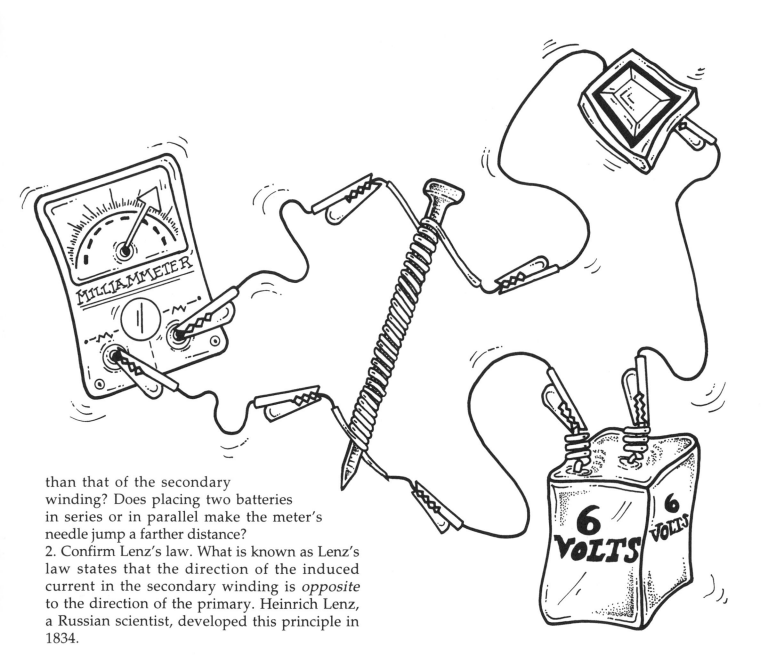

than that of the secondary winding? Does placing two batteries in series or in parallel make the meter's needle jump a farther distance?

2. Confirm Lenz's law. What is known as Lenz's law states that the direction of the induced current in the secondary winding is *opposite* to the direction of the primary. Heinrich Lenz, a Russian scientist, developed this principle in 1834.

# PROJECT 4–4
# Double Throw to Second Base
### Relays to remotely control other circuits

### YOU NEED

- 12-volt lantern battery
- dual "D"-cell battery holder
- two 1.5-volt alkaline "D"-cell batteries
- single-pole, single-throw switch
- two light-emitting diodes
- double-pole, double-throw, 12-volt relay
- insulated jumper leads with alligator clips on each end

requires a large current flow, or high voltage. This makes such a system safer. If only a little voltage is needed to turn the relay on, a switch, or button, that a human must come in contact with to activate the larger system would present no shock hazard. It would only have to hold enough current to turn on a relay. The contacts on the relay, however, can be designed to handle much larger amounts of current and voltage, thus the user is indirectly controlling the more dangerous circuit safely. A good example of this kind of relay system is a doorbell.

Since the contacts on the relay can be normally open or normally closed when the relay is at rest, relays can be used to complete one

A relay is an electrical device that allows current flowing in one circuit to control the current flowing in another circuit. The relay consists of a coil with an iron core, forming an electromagnet. A metalic plate is placed in position over the electromagnet and kept at a distance from it by a spring.

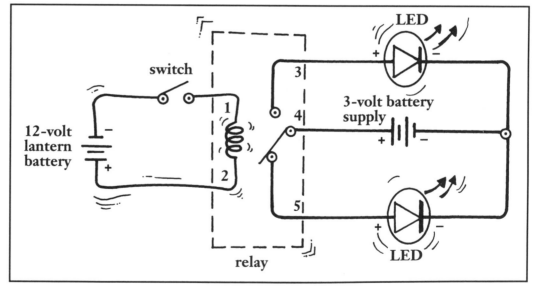

Connected to this plate are one or more switch contacts. When the electromagnet is at rest, some of the switch contacts may be touching (making a "closed" circuit) and some may not (making an "open" circuit). When the electromagnet is energized, those switch contacts that were open now become closed, and vice versa.

Relays are often used in situations where a small current flow can turn on a machine that

circuit when the electromagnet is off and another when the electromagnet is energized. As you know, you can connect a battery to a switch and a small lightbulb and the bulb will light when you turn on the switch to close (complete) the circuit. But what if you wanted to turn a lightbulb on when the switch was in the open position? A relay can be used to perform this task. This application is used in industrial electronics as well as in some remote-controlled TV sets.

Hypothesize that, by opening a switch, you can close another circuit through the use of a relay.

Assemble the circuit shown in the illustrations. Two power sources, in this case two separate battery supplies, are required for this circuit. The 12-volt battery supply delivers power to the relay coil when the switch is turned on. The 3-volt battery supply, made up of two 1.5-volt batteries in series, is used to light the LEDs (light-emitting diodes). Which of the two LEDs are illuminated depends on whether the relay's coil is powered or not. Notice that the two battery-powered circuits are electrically independent of each other.

One LED should be lit when the switch in the relay coil circuit is off, and the other LED should be lit when the switch is on. Label the switches to indicate whether the switch is on or off. Reach a conclusion regarding your hypothesis.

### Something more...

What applications, or uses, can you envision for this type of circuit? How does a thermostat in your living room control the furnace in your basement?

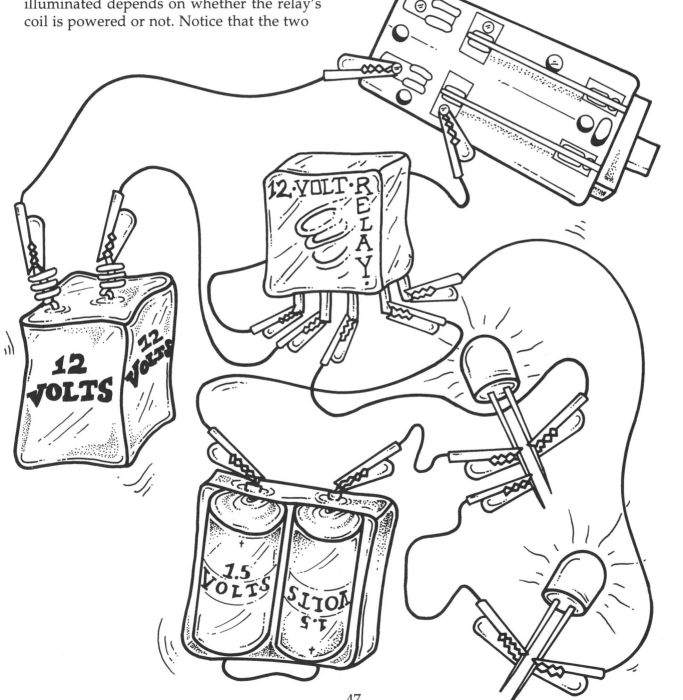

# PROJECT 4–5
# Motoring Your Generator
### Motors are the opposite of generators

**YOU NEED**

- two identical, 3-volt DC, miniature hobby motors, with small gear on shaft

- two "D"-cell alkaline batteries

- dual "D"-cell battery holder

- sensitive DC milliammeter

- insulated jumper leads with alligator clips on each end

- small piece of wood, screws, and hardware to mount motors

When an electrical conductor passes through a magnetic field, an electric current is induced in the conductor. The opposite is also true. A conductor carrying an electric current creates a magnetic field around it.

An electric motor consists of a magnet and an "armature," which is a coil of wire mounted on a shaft that rotates inside the magnet's magnetic field. A generator and a motor are similar in construction, but opposite in purpose. In a motor, electricity is put in and the armature turns, yielding a mechanical force. In a generator, a mechanical force is used to turn the armature and the output is electricity.

Hypothesize that, when the armatures of two identical motors are connected together and electricity is supplied to one, one motor will turn the electrical energy into mechanical energy and the second motor will turn mechanical energy into electrical energy. (An electric motor is the opposite of a generator.)

On a small piece of wood, mount two 3-volt DC miniature hobby motors facing each other. Use screws or other mounting hardware to secure the two motors firmly to the wood. The gears on the motor shafts must mesh so that

when the shaft of one motor turns the other will also turn.

Connect two 1.5-volt "D"-cell batteries in series, creating a 3-volt power source. Connect the power to one of the motors. Determine if electricity is coming out of the second motor by connecting a sensitive DC-milliampmeter to the generator.

### Something more...

1. Will the energy conversion become more efficient if more voltage is applied? Efficiency is equal to voltage out divided by voltage in. Do this computation for several different voltage inputs.

2. Can friction be reduced by lubricating the gears? Is it measurable?

# SECTION FIVE
# RESISTANCE AND CAPACITANCE

**R**esistors and capacitors are perhaps the most common passive components used in electronic circuits. They perform dozens of tasks, depending on how they are used in a circuit.

As the name implies, a resistor "resists" the flow of electrons, making their travel difficult. Resistors can limit the flow of current and drop voltage down to a desired level. The amount of resistance a resistor has is measured in "ohms."

Capacitors are storage devices. They consist of two metal plates separated by an insulating material, called a "dielectric." They do not generate electricity but they can store a small amount of it. Capacitors are used in many ways such as filters, temporary storage devices, and as blocks to DC (direct current) flow while passing AC (alternating current) signals. The amount of capacitance a capacitor has is measured in "farads." The most commonly used values, however, are only in millionths of a farad, called "microfarads."

# PROJECT 5–1
# Wired to Melt
## *Resistance in wire can produce heat*

---

### YOU NEED

- a 6-inch piece of thin nichrome wire (from a discarded toaster)
- two alkaline "D"-cell batteries
- dual "D"-cell battery holder
- insulated jumper leads with alligator clips at each end
- candle
- matches
- a piece of scrap paper
- two sections of 2 × 4 wood about 4 or 5 inches long
- two nails about 2 inches long
- hammer
- wire cutters

---

Often, heat is an undesirable side effect of electricity passing through a wire, for example in motors or such electronic appliances as TVs and stereos. But there are times when we can put the heat coming from electric current traveling in a wire to good use.

We have all seen the wires inside a kitchen toaster glowing orange as we make breakfast. The heating-element wire inside a toaster is made of metal wire called nichrome, an alloy of nickel, iron, and chromium. It has a high electrical resistance and warms up quickly. If DC (direct current) were applied to a piece of nichrome wire, would the part of the wire closest to the negative pole of the battery heat up before the side opposite? The negative pole of a battery supplies the electrons that flow through

the wire as they travel toward the positive pole, which is lacking electrons. Hypothesize that when DC voltage is applied to a piece of nichrome wire, the end of the wire closest to the negative source of the battery will heat faster than the other side.

Lay two pieces of 2 × 4 board on a workbench or old table so that the widest part is lying flat. Hammer a nail partway into each, leaving about a half inch to an inch of the nail extending above the board. These will be terminal posts, where the nichrome wire can be wrapped around and alligator clip leads from the batteries can be connected.

Cut a 6-inch piece of nichrome wire. Request some wire from your school's science teacher or ask an adult help you remove a piece from an old, discarded toaster. (An old toaster might be found at a yard sale for a dollar or two.)

Wrap one end of the nichrome wire around the nail in one of the pieces of wood. Wrap the other end around the nail in the other piece of wood. Gently push the two pieces of wood away from each other until the nichrome wire is slightly taut, so that the wire is straight and level.

Place a piece of scrap paper on the table under the nichrome wire to catch liquid wax. Then, using a lit candle, drip a thin coat of wax onto the whole length of the wire.

Use an insulated lead with alligator clips to connect the positive terminal of a dual "D"-cell battery holder to one of the nail terminal posts. Take the other clip lead and connect one end to the negative battery terminal. Clip the other end to the remaining nail terminal and observe the wax on the ends of the nichrome wire. Does the wax near the negative nail terminal begin to melt before the wax on the opposite end?

To check your project for validity, light and use the lit candle again to recoat the nichrome wire evenly with wax and repeat the procedure, but this time reverse the positive and negative clip leads. The wax will always melt first by the negative clip lead. Reach a conclusion about your hypothesis.

### Something more...

1. How do kinks in the wire affect heating? Tie a knot in the wire and repeat the experiment. Does the knot heat faster, slower, or the same as the wire near the negative battery terminal?

2. Cut an "H" shape in a piece of wood and suspend a piece of nichrome wire across the top two legs in the "H". Use clip leads for connection to a battery and use this tool as a candle cutter or a Styrofoam cutter.

3. Place a DC ammeter in series with the nichrome wire. Hypothesize that when current is first applied to the wire and the wire is cold, its resistance is low, and there will be a great amount of current flow. But as the wire begins to heat, its resistance increases and the current flow will decrease slightly.

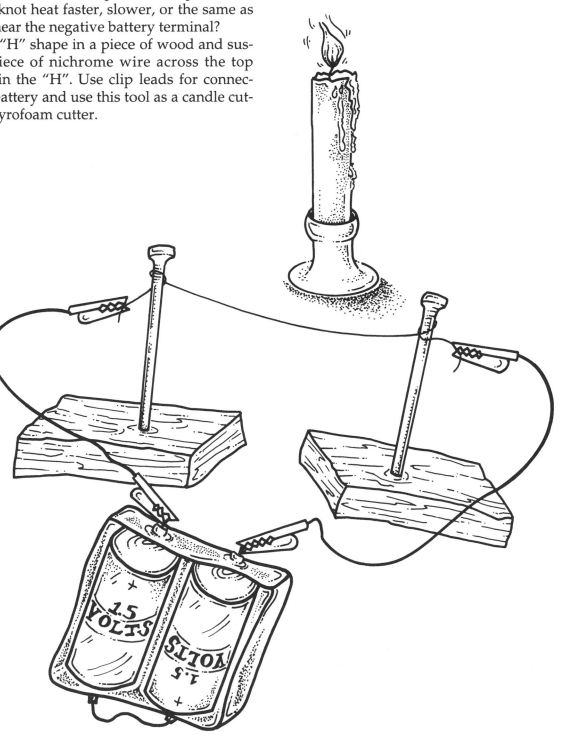

# Two Kinds of Sharing

*Comparing resistors in series and parallel*

---

### YOU NEED

- ohmmeter
- insulated jumper leads with alligator clips on each end
- four 100-ohm resistors

---

Electric current flows through a path of wire and other components that make up an electric circuit. In a circuit powered by a battery, the negative terminal on the battery is the source of the electrons that will work their way through the circuit to get to the positive terminal of the battery.

A resistor is an electrical component that "resists" or cuts down the amount of electrons that flow through it. Imagine driving a car at a fast highway speed on a smooth road and then suddenly hitting a rough and bumpy section of roadway. You would have to reduce your speed. Similarly, a resistor gives electrons a tough time and reduces the current flow.

Resistors come in different amounts of resistance, which is measured in units called "ohms." A low ohm value indicates that the resistor presents only a little hindrance to the electrons while a high number severely limits current flow.

Resistors can be placed either in series (in line with) or parallel (across each other) in a circuit. Hypothesize that resistors placed in series will make the total resistance of the circuit equal to the sum of the resistances, but resistors placed in parallel will share current flow. Therefore, the total resistance of a parallel circuit will be less than the lowest value resistor.

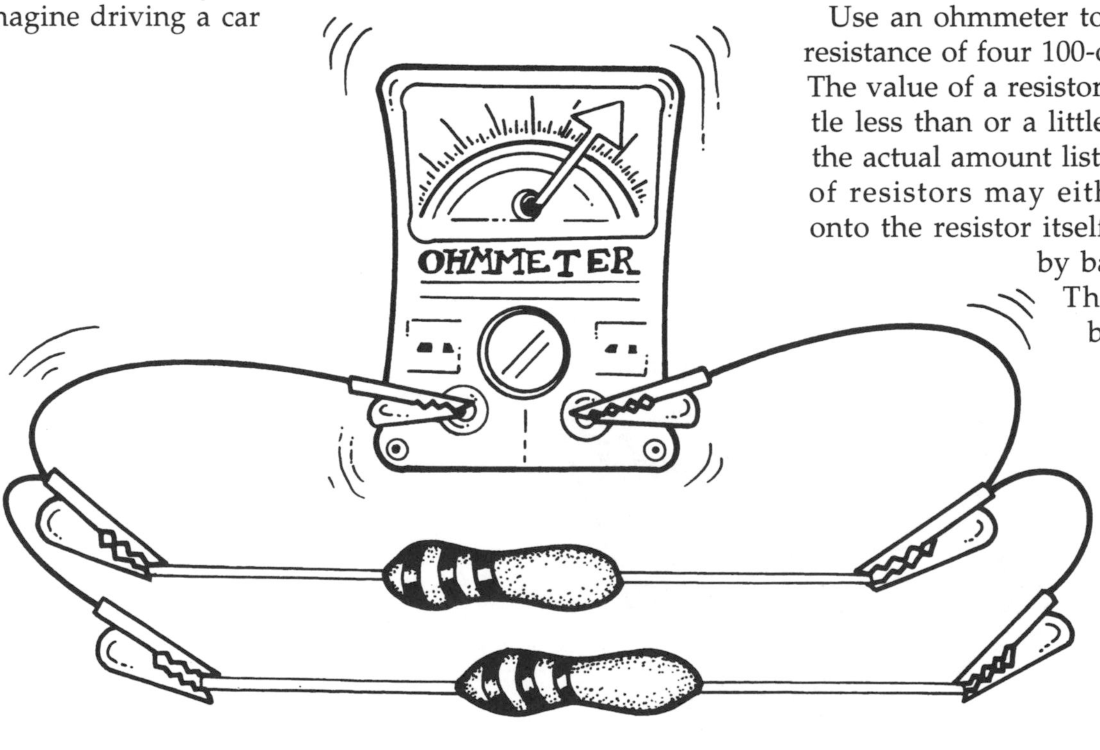

Use an ohmmeter to measure the resistance of four 100-ohm resistors. The value of a resistor may be a little less than or a little greater than the actual amount listed. The value of resistors may either be typed onto the resistor itself or indicated by bands of color. The first three bands indicate the resistor's value in ohms. If there is no forth band, then its accuracy is plus or minus 20 percent. That would mean that a 100-ohm resistor could really be as little as 80 ohms or as much as 120 ohms. A silver forth band means 10 percent "tolerance" (accuracy), and a gold band means 5 percent. Record the actual value of each resistor.

Twist one end of a 100-ohm resistor with one end of another. Using insulated jumper wires with alligator clip ends, connect the two 100-ohm resistors in series. Put the probes of an ohmmeter on the other ends of the resistors. Record the reading on the ohmmeter. Is the measured resistance the same as if you added the individual resistances of the two resistors together?

Twist both ends of two resistors together to put them in parallel (across) each other. Using insulated jumper wires with alligator clip ends, connect an ohmmeter. Record the circuit's resistance as measured by the meter. Is the resistance less than the lowest individual resistor value? (NOTE: When two resistors of the same value are in parallel, the total circuit resistance will be half the value of the resistors. In this case, ½ of 100

ohms is 50 ohms.) Reach a conclusion about your hypothesis.

**Something more**...

1. There is a formula for determining the total resistance for two resistors placed in parallel:
$$R \text{ total} = (R1 \times R2) / (R1 + R2)$$
For example, if the first resistor is 100 ohms and the second is 50 ohms, the total resistance would be:
$$R \text{ total} = (100 \times 50) / (100 + 50)$$
$$R \text{ total} = 5000 / 150$$
$$R \text{ total} = 33.3 \text{ ohms}$$
If there are more than two resistors, the formula is:
$$R \text{ total} = 1 / (1/R1 + 1/R2 + 1/R3...)$$

2. A potentiometer is a variable resistor, as is the volume control on your TV set. Use a potentiometer in parallel with a known resistor and measure the total resistance. Hypothesize that you can indirectly determine the resistance of the potentiometer in any position by measuring the circuit's total resistance and using the formula to calculate the unknown value.

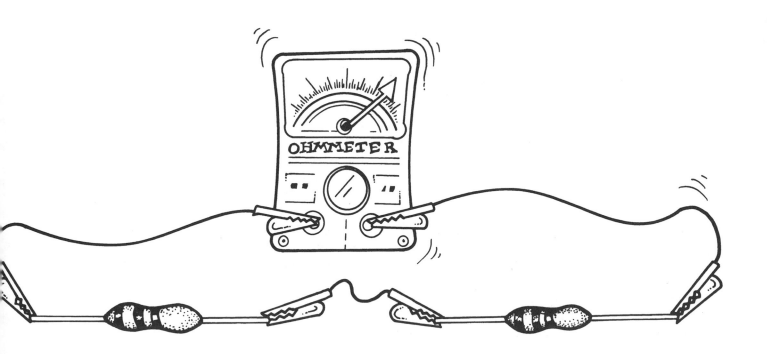

# PROJECT 5-3
# Brown Out
*Power loss in electric transmission lines*

## YOU NEED

- two flashlight bulbs
- two bulb sockets
- two 1.5-volt "D" cell alkaline batteries
- dual "D"-cell battery holder
- insulated jumper leads with alligator clips at each end
- 3-foot-long piece of nichrome wire
- three or four scale model telephone poles

A tremendous amount of electricity that power plants generate is wasted by resistance in the transmission wires that carry the electricity from the plant to our homes. While the resistance of the wires is very low, the total resistance in thousands of miles of wire becomes very significant. When Thomas Edison wired a town to bring electric lights to every home, he noticed that the lights were brighter near the power plant than those across town from it. The resistance in the long runs of wire reduced the amount of power available to homes located far from the plant.

Today, improvements have been made that reduce wasted electricity in transmission wires, but significant losses still remain a problem. Power companies use AC (alternating current) instead of DC (direct current). Transformers can be used with alternating current to step voltage up or down. As current travels through a wire, resistance creates a power loss in the form of heat. The amount of power wasted is determined by the following formula:

$$power = current^2 \times resistance$$

As you can see, if either the current or the resistance figures can be reduced, then the power consumed will be reduced. Also, because

$$power = voltage \times current$$

the same amount of power can be carried in the wires by either a high voltage with a low current or a low voltage with a high current. The power formula above suggests making current low to reduce wasted energy as heat. Therefore, power companies use transformers to step up the voltage at the power plant, thus reducing the current traveling in the wires. The voltage can then be stepped down by another transformer once it reaches your home, and higher current will be available for your use.

To further minimize resistance loss in wire, the diameter (thickness) of the wire can be increased. This works up to a certain point, after which further increasing offers no improvement. Imagine water rapidly flowing through a small hose compared to a larger hose. There is less resistance to the flow of water in the larger hose.

This project will demonstrate the concept of wasted power by transmission line resistance. It uses high resistance wire to simulate miles of transmission lines. Hypothesize that, because wire has resistance, electric power is wasted by the wires that carry it from the power plant to our homes.

Gather the materials listed above. Nichrome wire can be purchased through a scientific supply house or obtained from a discarded toaster. (You may be able to pick up an old toaster for a dollar or two at a yard sale.) Have an adult help you remove the wire, which is toaster's heating element. Three or four scale model telephone poles may be obtained from a hobby shop carrying model-train supplies. If you choose, use balsa wood and construct your own telephone poles.

# SECTION 5: RESISTANCE AND CAPACITANCE

Drape the nichrome wire along the scale model telephone poles. Use alligator clips to connect all components in the circuit, as shown in the illustration. One lightbulb will be connected near the battery supply and another three feet away, at the end of the nichrome wire. Which light is brightest? Reach a conclusion about your hypothesis.

**Something more...**

1. Use an ohmmeter to check the resistance of the wire when it is cold and again immediately after the bulbs have been operating for several minutes.

2. If the nichrome wire is kept cooled (run it through ice cubes), does the bulb at the far end appear any brighter?

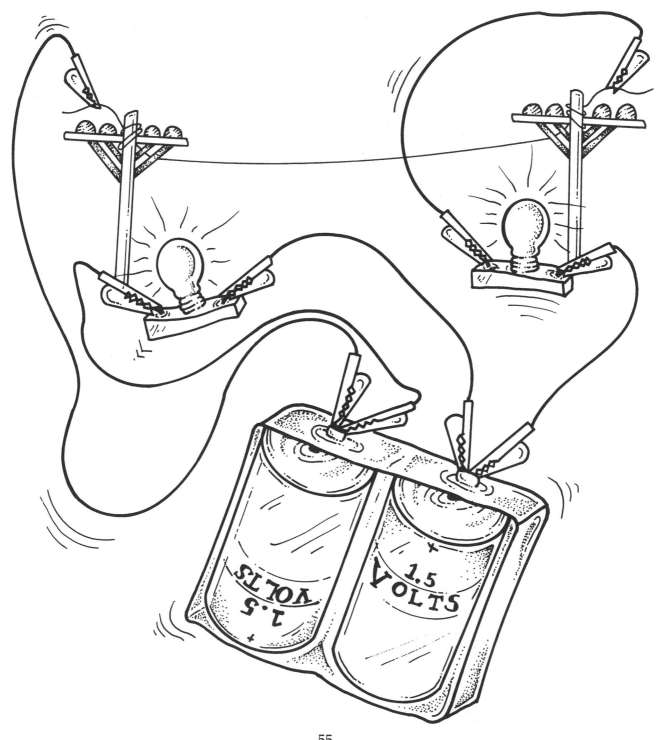

# PROJECT 5–4
# Secret Resistance
### Determining resistance by indirect measurement and computation

## YOU NEED

- flashlight bulb
- bulb socket
- two 1.5-volt "D"-cell alkaline batteries
- dual "D"-cell battery holder
- insulated jumper leads with alligator clips at each end
- DC ammeter

It was in 1827 that Georg Ohm, a German high school teacher, discovered a very simple relationship, but one of the most valuable in the field of electricity. This relationship was between resistance, voltage, and current. Through experimentation, he discovered that the current flowing in a conducting device was directly proportional to the voltage across it. He developed a formula which is now known as "Ohm's law": voltage = current × resistance, where voltage is expressed in volts, current in amperes and resistance in ohms. People in the field of electronics remember the formula as

VOLTS

AMPS × OHMS

If two of the variables are known, the third can be determined by placing your hand over the variable you want to know and covering it. For example, to determine the voltage if the current and resistance is known, place your hand over VOLTS and you get the formula AMPS × OHMS. To find the amps (amperes), the formula becomes volts/ohms.

What is the resistance of a flashlight bulb when it is lit? An ohmmeter can only be used to read the resistance across the bulb when it is off and the filament is cold. But in order to light the bulb, voltage must be applied to it. An ohmmeter cannot read the resistance of any device that has power applied to it. So how can the resistance of the hot bulb be determined?

Hypothesize that, by the use of indirect measurement and Ohm's law, the resistance of a hot, illuminated flashlight bulb can be determined.

Construct the circuit as shown in the illustration. Record the reading on the DC ammeter. The voltage across the bulb is 2 × 1.5 volts (or 3.0 volts), since there are two batteries in series and their voltages add together. Using Ohm's law, the bulb's resistance is equal to the voltage (3.0) divided by the ampere reading on the meter.

## Something more...

What is the difference in resistance between the cold bulb and the bulb when the filament is hot?

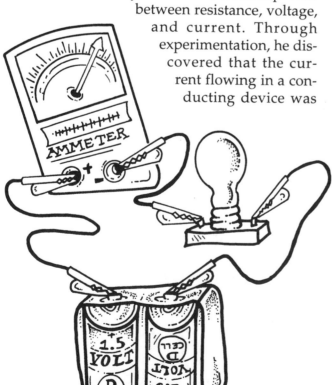

# PROJECT 5–5
# Red Hot Stop
## Thermistor resistors

Most types of metals have very low resistance to the flow of electrons when their temperature is cool or cold. There are some metals, however, which behave much differently. These materials are called "semiconductors" because at low temperatures they act like insulators but at higher temperatures they pass more current.

Some kinds of semiconductors are used in the making of diodes, transistors, and integrated circuits. Other semiconductor materials, such as uranium oxide, nickel-manganese oxide, and silver sulfide, are used to make "thermistors," which are temperature-sensitive conductors. Such a device could be used in a circuit to detect an increase in temperature, perhaps tripping a relay to sound an alarm if a food-freezer fails and temperatures inside the freezer are rising to the point where the food would spoil. In fact, thermistors are often used in industry in various temperature-measurement applications.

In this project, we will construct a circuit that places a battery, a thermistor, and an LED (light emitting diode) in series. Hypothesize that at room temperature the thermistor's high resistance prevents the LED from lighting brightly, but as its temperature is raised, the resistance becomes lower and the LED will increase greatly in brightness.

Construct a circuit placing two 1.5-volt batteries in series with a thermistor and an LED. Use jumper wires with alligator clips on them to connect the components together. The LED will light. Have an adult light a candle and secure it in a candle holder. Holding the thermistor by the alligator clip leads, move it close to the candle's flame. Do not put the thermistor directly into the flame. Allow a few seconds for the thermistor to change temperature. Does the LED light brighter? Was your hypothesis correct?

## YOU NEED

- LED (light emitting diode)
- thermistor
- insulated jumper leads with alligator clips on both ends
- dual "D"-cell battery holder
- two 1.5-volt "D"-cell batteries
- candle and candle holder
- an adult with matches

## Something more...

1. How long did it take for the thermistor to respond to the change in heat?
2. Compare the resistance and temperature of a piece of nichrome wire to that of a thermistor. Draw a chart illustrating resistance versus temperature for both.
3. What is the "threshold" voltage of the LED (the point at which it has enough voltage to turn on or off)?

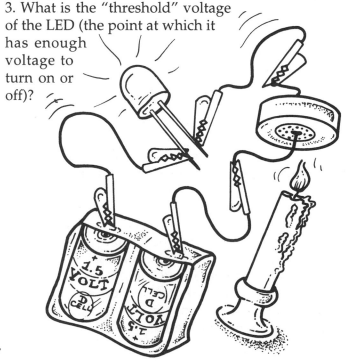

# PROJECT 5–6
# Divide and Conquer
### *Voltage divider circuits*

### YOU NEED

- 10k ohm resistor
- 10k ohm potentiometer
- DC ammeter
- dual "D"-cell battery holder
- two 1.5-volt "D"-cell batteries
- insulated jumper leads with alligator clips at each end
- DC voltmeter (able to accurately show ¼ volts or less)
- ohmmeter

When several resistors are placed in series with each other, each resistor "drops" a part of the total "supply" voltage across it. For example, suppose a 12-volt battery was connected across two resistors of equal value that were in series. Then each resistor shares in dropping a portion of the total voltage across it. The diagram shows the circuit. The total voltage is 12 volts. If both resistors are equal in their value of resistance, then the voltage measured from Point A to Point B will be 6 volts, and the voltage measured from Point B to Point C will also be 6 volts. The total voltage across the resistors is, of course, 12 volts. In a series circuit, the sum of the voltage dropped across each resistor is equal to the supply voltage. This resistor arrangement is called a "voltage divider."

A "potentiometer" is a special type of resistor whose resistance can be varied by moving a wiper arm across a resistive material, such as carbon. Whether you are aware of it or not, you are quite familiar with potentiometers and their use. Volume controls on your TV and stereo are potentiometers whose resistance is changed by turning a knob attached to a shaft. Other controls on TVs and stereos, such as color, brightness, treble, and bass, are also potentiometers.

Hypothesize that, by using the concept of a voltage divider, the resistance of a potentiometer can be determined by measuring the voltage drop across the potentiometer, measuring the current flowing in the circuit, and then mathematically applying Ohm's law.

Construct the circuit illustrated. Potentiometers have three lugs (connection points) on them as shown. Only two are needed here. Use the center lug and one of the end lugs (it doesn't matter which one) in this project. Turn the shaft on the potentiometer to approximately the middle of its range. Using a DC voltmeter, measure the voltage drop across the potentiometer. As seen in the illustration, this would be measured between Point A and Point B. Record the voltage registering and the reading on the DC ammeter.

Using Ohm's law, where resistance in ohms = volts/amps, calculate the resistance of the potentiometer at its present shaft position.

Remove the potentiometer from the circuit. Without disturbing the shaft position, place the

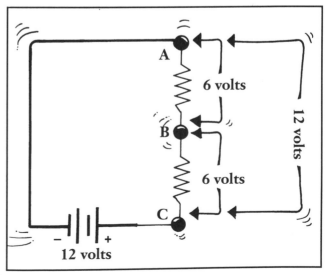

probes of an ohmmeter across the two lugs that were previously connected to the circuit. Measure the resistance. Is the measured resistance about the same as you calculated? Was your hypothesis correct?

**Something more...**

1. Confirm that the sum of both voltage drops is equal to the supply voltage by measuring the voltage drop across both resistors then adding the two readings together.

2. What happens if there are three resistors in series? Can you still determine the resistance of each one by knowing the current flow and the voltage drop across each one?

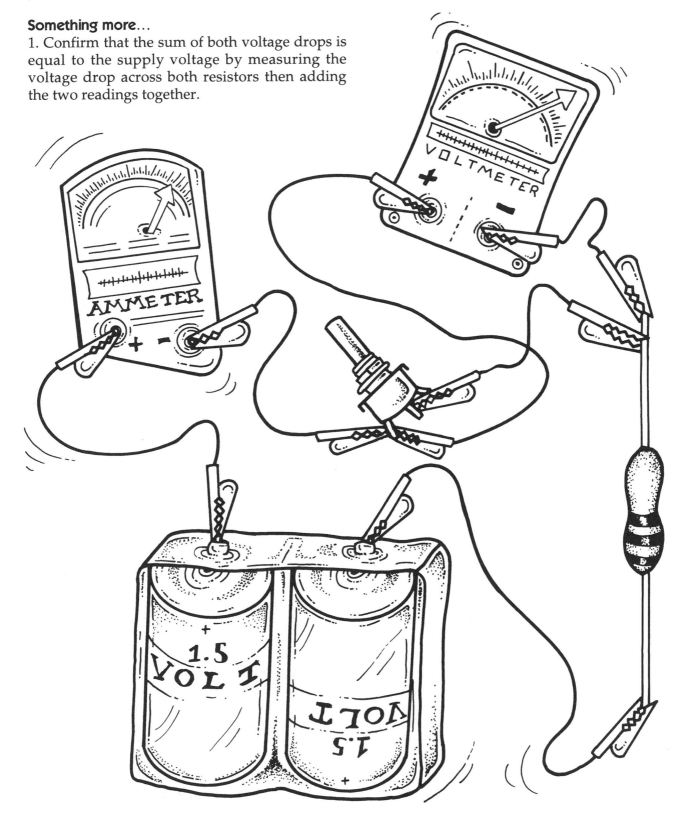

## PROJECT 5–7
# Charge It!
### *Charging time in a resistive/capacitive circuit*

---

### YOU NEED

- 9-volt battery
- 9-volt-battery clip connector
- two 220-microfarad, 35-volt capacitors
- voltmeter
- insulated jumper leads with alligator clips at each end
- 10k ½-watt resistor
- switch (such as: slide, toggle, or knife)
- stopwatch or watch with a second hand
- a friend

---

One of the factors that determines the electrical-storage capability of a capacitor is the surface area of the two plates that make up the capacitor. When voltage is placed across a capacitor that has no previous charge stored in it, electrons rush from the negative terminal of the power source (such as a battery) to one plate in the capacitor, while electrons on the other plate rush to the positive terminal of the power source. The electrons continue to flow until the voltage across the capacitor approaches the voltage supplied by the power source.

If a resistor is placed in series between the power source and the capacitor, the time it takes to charge the capacitor increases. This is called an "RC" circuit, meaning a "resistive/capacitive" circuit. Interestingly, electrical engineers have determined that the time it takes a capacitor in this kind of a circuit to charge to 63.2 percent of the power source voltage can be calculated by multiplying the value of the resistor (expressed in megohms) by the value of the capacitor (expressed in microfarads). This

charging time, expressed in seconds, is called a "time constant."

Hypothesize that the larger the value of capacitance in the resistive/capacitive circuit, the longer it will take to charge the capacitor (or capacitors) because of the increased overall plate surface area.

Construct an RC circuit as shown. Observe the polarity of the meter, battery, and capacitor when connecting them.

Here is where a friend will be needed. When the on/off switch is closed, the voltage indicated on the meter will slowly rise. Have a friend begin a stopwatch the moment you close the switch and call out each second. Quickly, write down the voltage reading at that moment. After a short time, the voltage will stop increasing and appear to level off.

Draw a graph, with time in seconds on the X axis and voltage on the Y axis, and plot the data collected.

Next, take an insulated jumper wire and connect it across the capacitor, holding it there for at least one minute. This discharges the capacitor.

Connect another 220-microfarad capacitor across the one already in the circuit, placing the two capacitors in parallel to each other. Repeat the charging and recording procedure, recording the time versus the voltage across the capacitor. Graph the data collected and reach a conclusion about your hypothesis.

### Something more...
1. Place a milliammeter in series with the resistor and capacitor. When the switch is closed, observe what happens to the voltmeter reading and the milliammeter reading. The capacitor acts as a direct short the moment the switch is closed. All voltage is dropped across the resistor. Current leads the voltage when the capacitor is charging. On graph paper, chart both the

charging current and the voltage with respect to time.

2. What happens to the charging time of the capacitor if the value of the resistor is increased, decreased? Try values of 100 ohms, 1k (1,000 ohms), 100k, 1meg (1 million ohms).

3. Try placing different values of resistors across the capacitor to lessen the discharge time.

4. What happens to the charging time when two capacitors are placed in series instead of parallel?

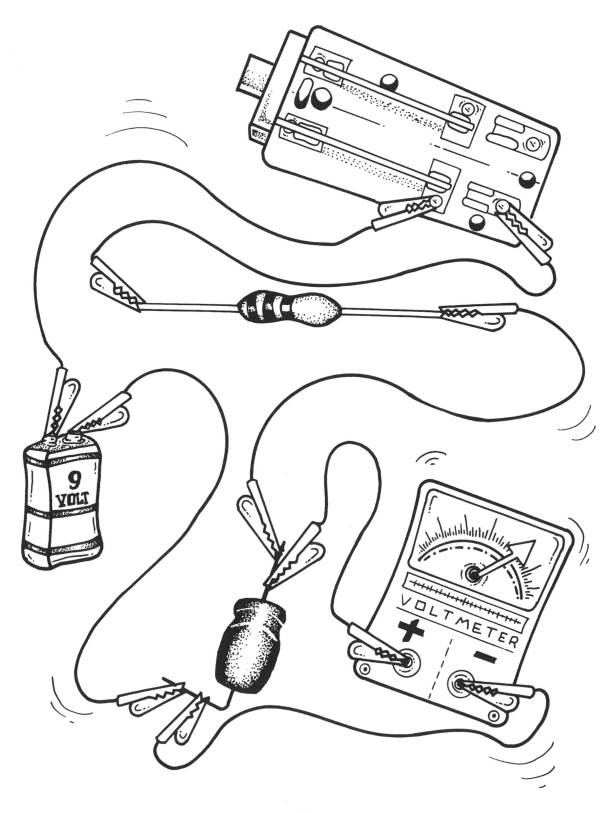

# SECTION SIX
# GENERATING ELECTRICITY

Human beings first used their own muscles to accomplish the tasks they needed to do. As time went on, various tools that could utilize animal power were invented. A poem written in 85 B.C. mentions the use of a waterwheel to harness the power of flowing water. The wind, too, has long been used to fill the sails of ships. Windmills were probably developed around the 700s in Persia.

Today, electricity is one of our main sources of energy, but it has to be generated and sent over transmission lines to our homes and businesses. Power utility companies generate electricity in several ways, including burning fossil fuels, harnessing moving water (hydroelectric generators), and nuclear energy.

Electricity can also be generated in other ways. In nature, lightning is a most powerful display of electricity. The electric eel, a fish that lives in the muddy waters of the Amazon River in South America, can generate shocks as strong as 300 volts, enough to stun a full-grown man.

Chemicals can be combined to form batteries that supply electric current. A thermocouple is a device made of two dissimilar metals which, when in contact with each other and heated, produces electricity. The electromotive force that is produced from a thermocouple, however, is in such small quantity that it is not used as a source of electricity, but as a temperature sensor in various devices. Solar cells convert sunlight to electrical potential. Piezoelectric energy is generated when crystals of certain materials are subjected to pressure, which makes it ideal for such applications as in microphones.

But, for supplying great amounts of electricity, the magneto concept is most widely used. This is where a coil cuts through a magnetic field, generating an electromotive force in the wire.

# PROJECT 6-1
# A Salt and Battery
*Determining peak power from wet-cell battery as load is applied*

## YOU NEED

- aluminum strip (about 1 inch wide by several inches tall)

- carbon strip (about 1 inch wide by several inches tall) or carbon rod

- water

- teaspoon of table salt

- lemon juice

- milliammeter (DC 50 milliamps full scale)

- glass beaker or wide-mouth peanut butter jar

- insulated jumper leads with alligator clips on each end

When a certain pair of dissimilar metals are suspended in an electrolyte (a liquid solution that conducts electricity), a voltage potential is generated across the metals (making a "wet-cell" battery).

Different combinations of metal electrodes and electrolyte solutions will yield different electrical results. Some combinations may not produce any electricity, while others are good producers. Are wet-cell batteries able to produce a steady current flow, or do they fluctuate? Does a wet-cell battery accumulate a charge over time? When a milliammeter is bridged across two electrodes (two dissimilar metals) suspended in an electrolyte solution, does the current flow build to a peak, or does it start at its peak output and fall off? Hypothesize that the peak current delivered by a wet-cell battery which uses salt water as an electrolyte will either build up or fall off but does not stay steady after the circuit is initially closed to allow current flow.

Place a strip of aluminum and a carbon rod in a glass beaker or wide-mouth peanut butter jar. Attach one end of an insulated jumper wire to the aluminum strip and one end of another jumper to the carbon rod. Fill the beaker with water and stir in a teaspoon of table salt. Attach one end of one of the jumpers to a terminal on a milliammeter. Observing the meter's needle, touch the other jumper to the other meter terminal. (NOTE: If the needle moves *backwards*, to below zero, reverse the two jumpers on the milliammeter terminals marked + and −. Since we are measuring direct current, we must consider polarity, which is the direction of current flow.)

Record the initial meter reading. After a few seconds, note and record the meter reading again. Continue to monitor the current flow, watching for a trend, steadily falling, rising, leveling off, and so on.

Repeat the procedure but this time replace the electrolyte solution (salt water) with a solution of lemon juice and water. Are the results different? Did both wet-cell batteries deliver a steady current flow, or did it quickly change once current began to flow? Reach a conclusion about your hypothesis.

You may wish to try other combinations of electrodes in salt water; such as zinc, lead, iron, and tin.

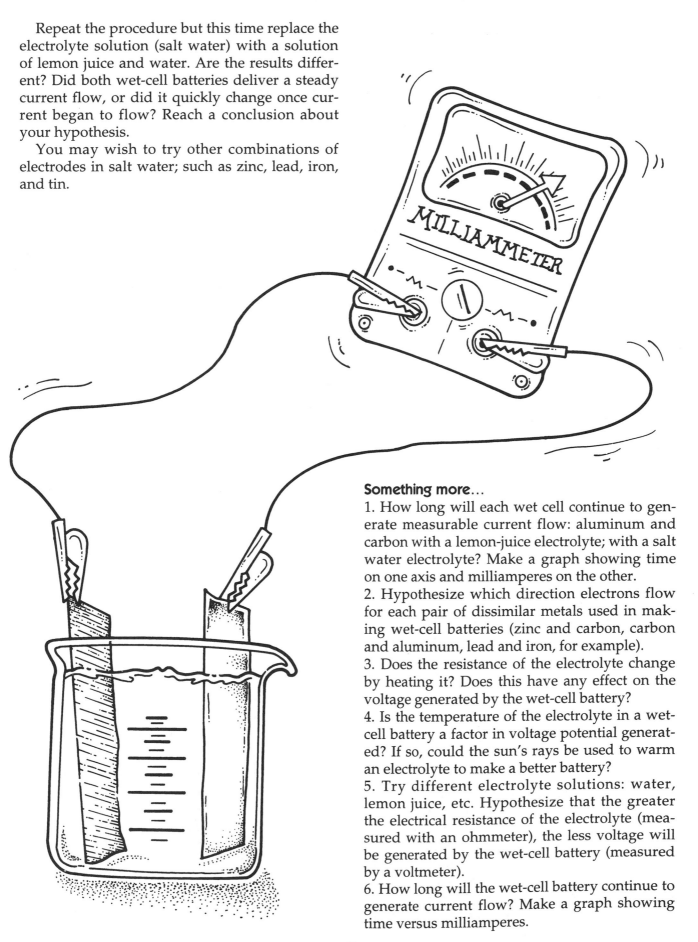

**Something more…**

1. How long will each wet cell continue to generate measurable current flow: aluminum and carbon with a lemon-juice electrolyte; with a salt water electrolyte? Make a graph showing time on one axis and milliamperes on the other.

2. Hypothesize which direction electrons flow for each pair of dissimilar metals used in making wet-cell batteries (zinc and carbon, carbon and aluminum, lead and iron, for example).

3. Does the resistance of the electrolyte change by heating it? Does this have any effect on the voltage generated by the wet-cell battery?

4. Is the temperature of the electrolyte in a wet-cell battery a factor in voltage potential generated? If so, could the sun's rays be used to warm an electrolyte to make a better battery?

5. Try different electrolyte solutions: water, lemon juice, etc. Hypothesize that the greater the electrical resistance of the electrolyte (measured with an ohmmeter), the less voltage will be generated by the wet-cell battery (measured by a voltmeter).

6. How long will the wet-cell battery continue to generate current flow? Make a graph showing time versus milliamperes.

# PROJECT 6-2
# Idaho Electric Company
### *Electrolytes in potatoes*


### YOU NEED

- one copper and one zinc strip (about 1 inch wide by 2 to 3 inches long)
- fresh potato
- sensitive DC milliammeter
- insulated jumper leads with alligator clips on each end
- sunny window

**B**et you thought french fries were only for eating! Do you have a problem seeing a potato as a power source? It is possible though, but quite a large number of potatoes would be needed to power even the smallest electrical device.

Electricity can be produced by chemical action between two dissimilar metals with an "electrolyte" substance in between. An electrolyte is a liquid solution that conducts electricity and is used with certain metals to generate electricity.

Since a potato is made up largely of water, a "potato battery" is actually a type of wet-cell battery.

What happens to the electric-generating ability of a potato as it gets old, loses moisture, and begins to shrivel up? Do you think it will lose the ability to produce as much electricity as when it was fresh, because there is less water? Or do you think that, since there is less water, the substances in the water that make it an electrolyte are more concentrated and the potato will be capable of creating *more* electricity? Form a hypothesis based on either of these theories.

Carefully cut a fresh potato in half and discard one section. Push a small copper strip into one side of the potato half and a small zinc strip in the other side. Use alligator clip leads to connect each metal strip to the terminals on a sensitive DC milliammeter. Measure and record the current indicated on the meter. Remove the metal strips.

Place the potato in a sunny window. At the end of each day, insert the metal strips and record the amount of current indicated on the meter. Continue to check the output of the potato battery daily for several weeks.

Compare the daily readings. Has the battery output risen, fallen, or remained the same over the several-week period? Reach a conclusion about your hypothesis. Do you think pH was a factor?

**Something more...**
1. Perform this experiment with different types of potatoes, such as red skin, sweet, even cooked french fries.
2. Measure the output from a potato battery. Then microwave the potato until it is cooked. Measure the output again. Compare the cooked and uncooked outputs.
3. Perform the battery experiment using electrolytes other than potatoes, such as apples, oranges, vegetables.
4. Is there a relationship between carbohydrate content and the ability of different potatoes to produce current?
5. Replace the copper and zinc strips in the potato battery with strips of other types of metals. Which combinations give the best results?

# PROJECT 6-3
# Lighten-Up
## Solar cells

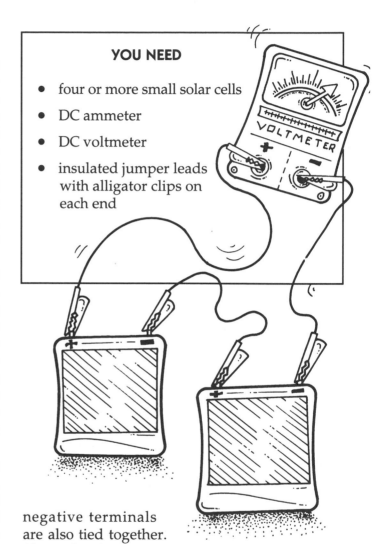

**YOU NEED**

- four or more small solar cells
- DC ammeter
- DC voltmeter
- insulated jumper leads with alligator clips on each end

The space age would have had a big setback had it not been for the invention of the "photovoltaic cell," commonly called a "solar cell." A solar cell is a semiconductor device that converts light into electricity. Satellites in orbit around the Earth depend on solar cells to convert sunlight into electricity to power the equipment on board. Solar cells are also useful when a small electrical source is needed in a remote area where no electric power lines exist; for example, an isolated weather station that transmits data to scientists at a distant location.

Often you will see small mushroom-shape lights outlining walks and driveways of homes. To avoid the need for burying wire and making connections between a power source and each lamp, some light sets are completely self-contained. They require no external power connection because solar cells on the top of the lamps collect sunlight during the day and convert it to electricity. Rechargeable batteries store the electricity and, when the sky is dark at night, power the lamps.

The problem with solar cells has been that a small cell produces very little electrical power. A small cell (measuring about 2 × 4 centimeters) may be capable of delivering only about 0.3 amperes at ½ volt in full sunlight. A typical two-cell flashlight, by comparison, requires about 3 volts and 450 milliamperes to yield a bright light. For solar cells to be truly put to work, their power output needs to be increased.

Hypothesize that solar cells can be arranged and connected in various ways (combinations of series and parallel) to produce a variety of voltage and current requirements and thus make them useful for powering many devices.

For series connections, the positive terminal of one cell is connected to the negative terminal of the next. For parallel connections, all the positive terminals are connected together and all the negative terminals are also tied together. Show that by connecting solar cells in series, the voltage is increased (use a voltmeter). Show that connecting solar cells in parallel increases the amount of current the cells are able to deliver (use a DC ammeter). Connect two sets of cells in series and then place both sets in parallel. Measure the total voltage and the total available current and reach a conclusion about your hypothesis.

You may wish to enhance your science fair display by having the solar-cell assemblies power light-emitting diodes, small bulbs, and miniature motors.

# PROJECT 6–4
# Niagara Sink
## *Hydroelectric power*

## YOU NEED

- dual "D"-cell battery holder
- sensitive DC voltmeter, capable of reading less than 1 volt
- insulated jumper leads with alligator clips on each end
- very thin balsa wood
- utility knife or wood cutting tool
- an adult
- sink with tap
- varnish or shellac
- paint brush
- glue

To generate electricity, barrier dams are built across rivers to direct the flowing water past huge, water-catching mechanisms that rotate the armatures of generators. These great turbines generate tremendous "hydroelectric power" (electricity from flowing water) which does not use up our natural resources, as does energy derived from burning oil and coal—resources that are not renewable. The faster the armature of a generator rotates, the more power it provides.

Hypothesize that increasing the velocity of water hitting paddle-wheel fins attached to a generator will increase the voltage output.

Have an adult with a utility knife, or other balsa-wood cutting tool, cut out a small disk about 1¼ inch in diameter. You may want to place something round, such as the top of a small jar, on the wood and trace it. Into the edge of the wheel, cut six equidistant slots about ¼ inch deep. Cut six small flat squares to act as the blades of a paddle wheel. Insert these blades into the disk's slots.

The wood should now be sealed to make it waterproof. Varnish or shellac the wheel assembly. When dry, make a small hole in the center of the wheel for the shaft of the motor (which will be our generator). Use glue to secure the wheel to the shaft. Connect the probes of a sensitive voltmeter (one capable of reading below 1 volt) to the generator. Hold the paddle-wheel device under a sink tap and turn the water on slightly. Watch the voltmeter readings as you slowly increase the amount of water coming from the tap. Does the voltage increase as you increase the velocity of the water? If so, your hypothesis was correct.

### Something more…
1. What happens if the water falls from higher up?
2. Research the Niagara Falls generating plant, Boulder Dam, and other hydroelectric plants.
3. Can you get your generator to produce enough power to illuminate a light-emitting diode (LED)?

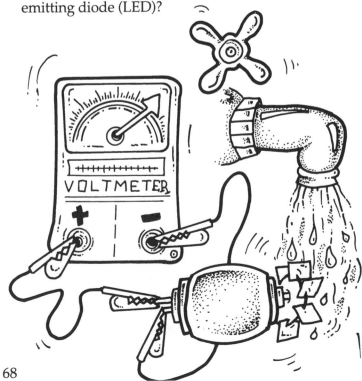

# PROJECT 6-5
# Threshold of Light
## *Using calculator's solar cells to measure light*

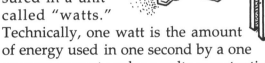

**I**ncandescent light bulbs, normally used in household lamps, are rated by the amount of electrical power they use. Electrical power is measured in a unit called "watts." Technically, one watt is the amount of energy used in one second by a one ampere current under a voltage potential of one volt. From experience you know that light bulbs with higher wattage ratings are brighter than those with lesser ratings. But is the wattage rating a true indication of the light an incandescent bulb will emit? Hypothesize that the light generated by a 100-watt bulb is about the same as two 50-watt bulbs (or choose to hypothesize the opposite is true).

To determine if the light from one 100-watt bulb is approximately equal to two 50-watt bulbs, some type of light-measuring device is needed. We will use a solar-powered calculator.

Place a lamp on a table in a dark room and remove the shade. Be sure the lamp is unplugged from the electrical outlet. Screw a 100-watt bulb into the lamp's socket. Plug the lamp in and turn it on. Except for the light from the lamp, the room should be dark.

Enter some numbers into a solar-powered calculator. Stand near the lamp and hold the calculator facing the lamp. Slowly move away from the lamp, as you watch the number display. At some point, the calculator will be so far from the bulb it will not receive enough light to display the numbers. Mark the spot where no numbers are seen and measure the distance from that position to the lamp.

Turn the lamp off and unplug it from the out-

| YOU NEED |
| --- |
| • 100-watt light bulb |
| • two 50-watt light bulbs |
| • two table lamps with removable lamp shades |
| • solar-powered calculator |
| • yardstick or tape measure |
| • dark room |

let. Place a second lamp next to the first. Both lamps should be about the same height, but need not be exactly the same. Put a 50-watt bulb in both lamps. Plug them in and turn them on. Again, holding the calculator, walk away from the lamps. Does the display disappear when you reach the marked point? Does it go out before you reach the mark, or not until you are well past it? Reach a conclusion about your hypothesis.

### Something more...

1. The project above assumes that all bulbs of the same wattage give the same light output. Check the lumens at fixed distances such as 1, 2, 3 and 4 feet from different lightbulb manufacturers. A photographer's light meter would be helpful.
2. Is there a mathematical relationship between the amount of available light at a certain distance from the bulb and the wattage of the bulb?
3. Use candles instead of light bulbs and repeat the above experiment. Does the same hypothesis hold true?
4. How much light is needed to do more than light the display but to actually operate the calculator correctly, or if the calculator gets enough power to light the display does it perform computations correctly?

# PROJECT 6-6
# Meter Made
*Periods of peak use of electric power in the home*

---

### YOU NEED

- access to your home's electric meter
- clock
- paper and pencil

---

We often take the convenience of electrical energy for granted. If we want to see in a dark room, we simply switch on the light and the room is filled with brightness. But using electricity costs money. Utility companies place an electric meter on each house to measure the amount of electricity used, so they know how much to bill. The utility companies measure the power used in "kilowatt-hours," one kilowatt being equal to 1,000 watts. It takes one kilowatt-hour of energy to operate ten 100-watt light bulbs for one hour.

Does your home use more energy during the day or the night? Consider that during the day, parents may be at work and children at school and no one may be home. If the house uses electric heat, sunshine entering windows during the day may reduce the need for electric heat, while the sunless, colder evenings may need more.

Choose two 12-hour periods, one during the day and one at night. The day hours, for example, may run from 8 a.m. to 8 p.m. and the night period from 8 p.m. to 8 a.m. Hypothesize in which period (day or night) most electric power is used in your home.

Electric meters have dials on their faces that are used to measure electric power consumption. Similar to the car odometer, which measures mileage, the numbers continue to count upward. The meter is not reset by the utility company "meter reader" who comes by each month.

To determine the amount of energy used over a given period of time, you need to record the number showing on the dials at the beginning and at the end of the time period and subtract to get the difference. The dials on an electric meter are read from right to left. From right to left, the dials indicate the ones, tens, hundreds, and thousands places.

In the morning, at 8 a.m. (or whatever time you have decided to use) read and record the number indicated by the dial on the electric meter. At 8 p.m., or twelve hours later, read and record the number again. Subtract the morning from evening number to get the the number of kilowatt-hours of energy used during the day.

At 8 a.m. the following morning, read and record the meter dials and subtract this number from the one recorded when you read the meter at 8 p.m. the night before. The result is the number of kilowatt-hours used during the twelve-hour night period. Compare the two numbers and reach a conclusion about your hypothesis.

### Something more...

1. How much does it cost for the electricity used in your home during a typical 24-hour period? From the monthly electric bill, you can find the cost per kilowatt-hour. Depending on where you live, this may be 5 cents to 15 cents per kilowatt-hour. Record the electric meter reading at 8 a.m. and again twelve hours later. Find the difference between the two numbers. Multiply this number by the cost per kilowatt-hour to determine the cost for the 24-hour period. Do this for 7 days and calculate the average cost per day. Which day of the week had the highest use of energy? Which day had the least? Why do you think the readings on those days differed?

2. Determine the cost to run each electrical appliance in your home for one hour. Make a list of each electrical device: TV, computer, toaster, refrigerator, hair dryer, iron, and others. Use the

stickers and tags on the back of the appliances or refer to the unit's owner's manual to find its power consumption. For example, if the utility company in your area charges 12 cents per kilowatt-hour and the sticker on the back of the TV set lists energy consumption as 125 watts (which is .125 kilowatts), then:

$$0.12 \times 0.125 = 0.015$$

or 1½ cents per hour. Watching TV for 4 hours, then, would cost

$$4 \text{ hours} \times 1½ \text{ cents} = 6 \text{ cents}$$

Which electrical appliances are the heaviest users of electric power? Which cost the most to operate over the period of one year? An iron may be a heavy user of electricity but if it is only used to iron clothes once a week, it may not be the most costly appliance to operate over the period of a year.

3. Can you reduce your home's monthly electric bill by practicing electrical conservation? What ways do you think you might be able to lower consumption and save kilowatt-hours?

electrical while you are observing the meter. Be sure the refrigerator compressor is off while observing. Compare the number of revolutions per minute at this point with the number observed at other times, such as when a toaster, clothes dryer, air conditioner, or steam iron is running. The faster the disc rotates, the more energy is being used.

5. Household safety: What appliances should not be plugged into the same receptacle and used at the same time?

4. Electric meters have a round disc that rotates as electricity is used. A mark on this disc and the use of a stopwatch, or watch with a second hand, can be used to determine how many revolutions per minute the disc is rotating (count them). Reduce the electrical consumption in your home to as low as possible. Ask everyone not to use anything

# SECTION SEVEN
# SOLID-STATE ELECTRONICS

In our homes are many work-saving appliances: washing machines, dryers, refrigerators, swimming pool pumps, vacuum cleaners. These are "electrical" appliances. Within the electrical field of study is a special branch called "electronics." Electronics is the science and practice of using special electrical components to perform tasks that are not possible with other electrical devices. The key devices in electronics are diodes, transistors, and integrated circuits, which make up a group called "semiconductors." Vacuum tubes also come under this category, but are seldom used today. The unique behavior of these electronic components makes possible radio, TV, long-distance phone service, computers, space travel, satellite communications, and advances in science, industry, and medicine.

The invention of the vacuum tube ushered in the age of electronics. But these devices were bulky, expensive, and inefficient. The next technological advance came in 1948 at Bell Telephone Laboratories with the invention of the transistor. Transistors performed the same electrical tasks as vacuum tubes, but they were inexpensive to make, were extremely small, required little energy, and could work in a wide range of temperatures. The era of miniaturization had begun.

As technology improved, scientists were able to integrate thousands of transistors into one electrical component called an "integrated circuit." These "integrated circuits,"sometimes affectionately referred to as "chips," make possible many of the appliances we enjoy today: computers, VCRs, microwave ovens, cordless phones, video games, and camcorders, to name a few.

# PROJECT 7–1
# One-Way Street
## Using LEDs to prove diodes establish directional current flow when AC source is supplied

---

### YOU NEED

- 9-volt AC transformer
- rectifier diode
- 1,000-ohm, ½-watt resistor
- 3 LEDs (light emitting diodes)
- low-wattage soldering tool and solder
- spool of solid conductor hookup wire
- wire cutters
- an adult, to solder and supervise

---

The diode is an extremely useful electronic component. In the early days of this industry, electronic tubes were used to perform the tasks that small semiconductor diodes do today. Semiconductors are materials that are classified between conductors and insulators. They exhibit unique electrical characteristics and are used in a wide variety of applications.

In electronic circuits, diodes can be used to convert AC (alternating current) to DC (direct current) through a process called "rectification," as well as to perform a number of other tasks. When the task is to change current from alternating to direct, the diode is sometimes called a "rectifier."

This project involves constructing a circuit that will use a diode to convert AC to DC. In AC circuits, the polarity of the power source is constantly changing. In the United States, the rate of polarity change in the electricity supplied to homes is 60 times per second, an amount determined by the power utility company. In DC circuits, the polarity remains constant, and current flows only in one direction. The "anode" is the positive terminal on the diode and the "cathode" is the negative. When a positive voltage is applied to the anode and a more-negative voltage on the cathode, then current flows through the diode.

Hypothesize that a diode will establish a single-directional flow (DC) when an alternating current (AC) source is applied.

Construct the circuit as shown in the illustration. This project requires some soldering. If you have never used a soldering tool before, ask someone experienced with soldering to show you the proper technique. You can then practice by twisting two pieces of bare hookup wire together and soldering them. CAUTION: The tip of a soldering tool gets very hot. Use the special stand made to safely hold the soldering tool when you aren't working with it. Place an old piece of cardboard or wood on your work area to catch any solder that drips. Be alert! Have an adult supervise. As soon as you are finished soldering, unplug it. Remember that, although it may not look it, the tip will stay quite hot for a few minutes after the plug is disconnected.

In the circuit to be constructed, the AC power transformer provides the AC source of voltage to the circuit, stepping down the higher voltage in the wall outlet to 9 volts. This voltage is not only safer to work with, but LEDs (light emitting diodes) only need a few volts to work. A resistor is also added to further reduce the voltage which powers the circuit.

LEDs are a special type of diode. Similar to other diodes, they only let current flow through them in one direction. But when the polarity is correct and current flows through them, they glow like small lightbulbs. Different-colored translucent plastic cases let the light shine out. The most popular colors are clear, red, yellow, and green.

In our circuit, an LED is placed on the AC side of the diode being used as a rectifier. This LED will detect the presence of alternating current. Every time the polarity changes favorably for the LED, it will light; and when the polarity is reversed, which happens many times per second, it will not light. Our eyes, however, cannot even detect the flickering, since the 60 cycles is so fast.

Next in line is a diode that will only let current flow when the polarity of the voltage reaching it though the resistor is positive. In this instance, the diode becomes a conductor. When the AC polarity swings in reverse, the diode will stop conducting and it will act like an open circuit.

To prove the hypothesis, two LEDs are placed on the output side of the rectifying diode. Note the polarity of these LEDs. In reference to their polarity, one is connected in a way opposite to the other. Remembering that the rectifying diode will only conduct when the positive part of the AC cycle reaches its anode, which LED do you think will light? If one of the LEDs on the output side of the diode lights and the other does not, your hypothesis is correct.

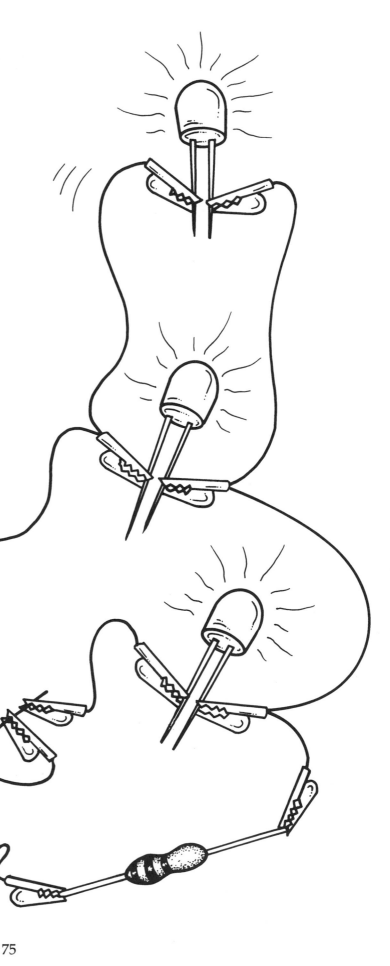

75

# PROJECT 7–2
# Cut the Sine in Half
### Speaker can detect diode's ability to change wave form

## YOU NEED

- speaker (preferably 6-inch diameter or larger)
- 9-volt AC transformer
- rectifier diode
- 1,000-ohm, ½ watt resistor
- low-wattage soldering tool and solder
- spool of solid conductor hookup wire
- insulated jumper leads with alligator clips on each end
- wire cutters
- an adult

The AC (alternating current) voltage supplied to our homes by the electric utility companies changes its polarity many times each second. The "frequency" of these changes is measured in "cycles" or "Hertz." In the United States, the frequency is 60 times per second, or 60 Hertz. The wave form that is created by this change is called a "sine wave." A sine wave shows graphically how the AC voltage changes polarity. The amplitude (size) of the voltage increases in one direction, then decreases and begins increasing in the opposite direction, completing the cycle by returning back to the starting point. The swing of a simple pendulum traces a sine wave.

By using a transformer, the higher AC voltage (120 volts in the United States) coming from an electrical wall outlet can be stepped down to a lower voltage. When a low-voltage AC sine wave is applied to a speaker, we hear a pure, low tone. When a diode is used as a rectifier to convert AC to DC, it changes the shape of the AC sine wave. Since the diode will only conduct when the voltage coming into it is of the correct polarity, the wave form coming out of the diode will be missing half of the sine wave each cycle. This will alter the sound we hear if that wave form is applied to a speaker.

Hypothesize that a diode changes the shape of an alternating current wave form and this can be detected by a change in sound.

Construct the circuit shown in the illustration. This project will require a little soldering. If this is new to you, ask someone with experience at soldering to show you the proper technique. You can practice by twisting two pieces of bare hookup wire together and soldering them. CAUTION: The tip of a soldering tool gets quite hot. Place the tool in a stand that holds it safely when you aren't soldering with it. An old piece of cardboard or wood placed on the work area will catch any solder that drips. Be very careful. In addition to doing soldering jobs, the hot soldering pencil tip can cause a painful burn. Have an adult supervise and, as soon as you are finished using the tool, pull the

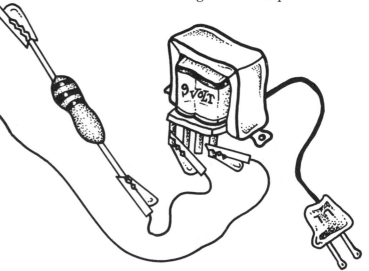

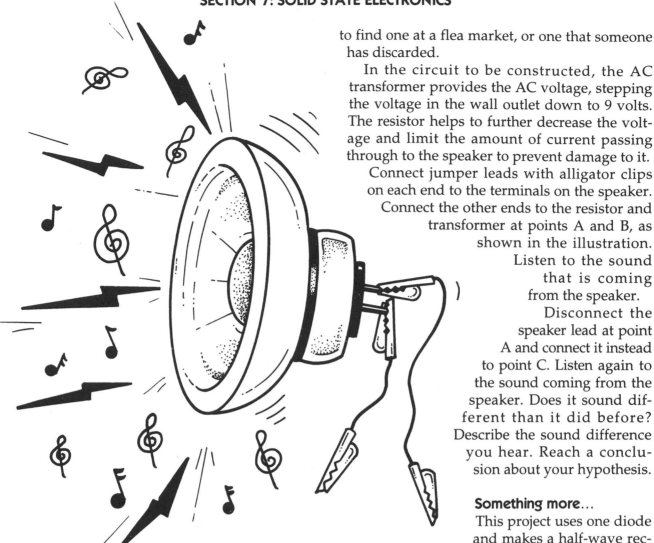

to find one at a flea market, or one that someone has discarded.

In the circuit to be constructed, the AC transformer provides the AC voltage, stepping the voltage in the wall outlet down to 9 volts. The resistor helps to further decrease the voltage and limit the amount of current passing through to the speaker to prevent damage to it.

Connect jumper leads with alligator clips on each end to the terminals on the speaker. Connect the other ends to the resistor and transformer at points A and B, as shown in the illustration. Listen to the sound that is coming from the speaker.

Disconnect the speaker lead at point A and connect it instead to point C. Listen again to the sound coming from the speaker. Does it sound different than it did before? Describe the sound difference you hear. Reach a conclusion about your hypothesis.

### Something more...

This project uses one diode and makes a half-wave rectifier circuit. Get a basic book on electronics and construct a full-wave rectifier using four diodes. Does the sound coming from the speaker appear any different?

plug. Be aware that the soldering tip, though it may not look hot, will stay hot for several minutes after the electric plug is disconnected from the wall outlet.

The speaker needed for this project is not critical, but the larger it is the better. We are dealing with a low frequency, and large speakers reproduce this lower frequency better than smaller speakers do. To save you money, a local TV-repair shop may be willing to sell you a used speaker from an old stereo, or you may be lucky enough

**Sine wave form**

# PROJECT 7-3
# Blinky
*Working with an integrated circuit oscillator*

### YOU NEED

- integrated circuit (such as Sylvania ECG-876) available through local TV repair shop)
- integrated circuit socket (8-pin, dual in-line)
- printed circuit mounting board
- 100-microfarad, 35-volt electrolytic capacitor
- 220-microfarad, 35-volt electrolytic capacitor
- 470-microfarad, 35-volt electrolytic capacitor
- LED (light-emitting diode), T-1 size
- spool of solid conductor hookup wire
- low-wattage soldering tool and solder
- "AA"alkaline battery
- "AA"-battery holder
- insulated jumper leads with alligator clips on each end

**E**arly electronic appliances were made up of discrete (individual) electronic components: tubes, resistors, capacitors, coils, and later, transistors. With the invention of the "integrated circuit," many discrete components could be "integrated," or built together, into one small package. An integrated circuit may contain thousands of transistors, diodes, resistors, and capacitors, all in one tiny package.

Integrated circuits are typically much smaller than the equivalent circuit would be using individual components, and they require much less electrical power. Integrated circuits come in various shapes and sizes. Since integrated circuits are destroyed by heat, sockets can be soldered onto printed circuit boards first and then the integrated circuits pushed into the sockets later. Use of a socket also allows easy removal and installation when replacing a bad integrated circuit.

The term "oscillator" describes a type of electronic circuit where the output from the circuit is a voltage that swings high and low or on and off. A pendulum on a cuckoo clock is a mechanical oscillator that swings back and forth.

In electronics, oscillators have many functions and are found in everything from radios to computers. This project uses an integrated circuit that contains components to make an oscillator circuit.

A critical component in an oscillator circuit is the capacitor. The value of the capacitor is crucial in determining the rate at which the oscillator will swing its output voltage. For this reason, the integrated circuit oscillator has pins to externally connect a capacitor. In this way the person using the integrated circuit can set the rate of oscillation by attaching the appropriate value capacitor.

Other external connections on the integrated circuit to be used include pins for a battery to power it and an output pin where an LED (light emitting diode) can be attached. We will connect an LED to the oscillator output to observe the rate of oscillation. As the output voltage swings high and low, the LED blinks on and off.

It is known that the value of a capacitor connected to the integrated-circuit oscillator will determine the rate of oscillation. Hypothesize that to increase the rate of oscillation, the value of the capacitor must be decreased (or you may hypothesize the opposite).

This project requires some soldering. Ask someone experienced with soldering to show

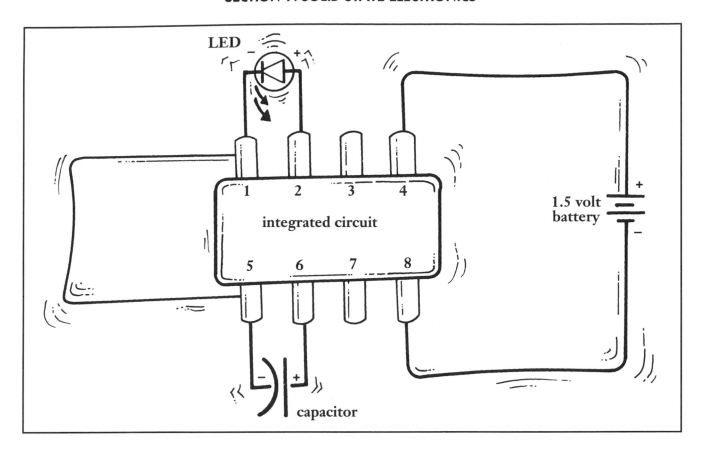

you the proper technique, if you haven't done it before. You can practice soldering by twisting two pieces of bare hookup wire together and soldering them.

CAUTION: A soldering tool tip gets quite hot. Use a stand to safely hold the pencil when you aren't soldering. Place an old piece of cardboard or wood under your soldering job to catch any drips. Remember that the tip of the soldering pencil can burn you. When you are finished using the tool, unplug it. Even so, be aware that the tip will stay hot for some minutes after the electric plug is disconnected, so do not touch it.

Construct the electronic circuit illustrated. During construction, note that the integrated circuit pins are numbered in a specific order. Integrated circuits typically have a dot marking pin 1. Also note that the polarity of capacitors, the LED, and the battery are important. Be sure to match + and – symbols, as in the illustration, to those on the components.

Do not solder any capacitors to the circuit board. Instead, strip ¼ inch of insulation off of both ends of two pieces of hookup wire. Each piece of wire should be about 6 inches long. Carefully, solder the end of one of the hookup wires to pin 1 on the integrated circuit socket and solder one end of the other piece of hookup wire to pin 2. Use these two free-standing wires to connect jumper leads with alligator clips. The clips will be used to easily attach different values of capacitors.

Once the circuit is connected, use the jumper leads and clip a 220-microfarad capacitor into the circuit. Observe the rate of flashing by the LED. Remove the capacitor and connect a 100-microfarad capacitor. Observe the rate of flashing. Remove the capacitor and connect a 470-microfarad capacitor. Observe the rate of flashing. Compare the flashing rates to the value of capacitance and reach a conclusion concerning your hypothesis.

**Something more...**
1. Use different values of capacitors. At what points of both increasing and decreasing capacitance is there no longer any visible change?
2. What value of capacitance will cause the LED to flash at a rate of 1 Hertz (Hz), that is, one flash per second? Capacitors can be placed in parallel with each other to increase the total circuit capacitance, which is equal to the sum of all the capacitors that are in parallel. How accurate is this flashing?

# PROJECT 7–4
# Yes or No
### AND and OR computer logic circuits

Many of the electronic circuits that make up a computer can be represented as single "blocks" that perform a handful of operations. Such blocks include storage of data, "flip-flops," and "gates." Inside the computer, these blocks actually consist of many electronic components: resistors, transistors, capacitors, and integrated circuits.

Just as a gate in a fence will let an animal through if it is open and will prevent them from passing through when it is closed, a "logic gate" refers to a circuit that will allow passage of a signal *under certain conditions*.

The two most common types of computer logic gates are the AND gate and the OR gate. Consider a cow pen that has two fence gates in series with each other. Let's represent a closed gate condition by the number zero, meaning no cows can pass. If one gate is open, the condition is represented by the number one. All information stored in a computer is in the form of binary numbers (base two), either a one or a zero. As you can see from the illustration, both gates must be open for the cows to pass out of the pen.

Computer logic gates have two or more "inputs" and one "output." The accompanying chart for AND and OR gates shows the status of the outputs for each possible input condition. Both inputs of an AND gate must have a 1 present in order for the output to be 1. Similarly, both the first fence gate and the second fence gate must be open in order for any cows to pass through.

The illustration on the opposite page shows how the OR gate relates to our cow-pen analogy. In the AND gate scenario, both gates have to be open before cows can pass. In the OR configura-

AND                                            OR

### Logic AND Gate

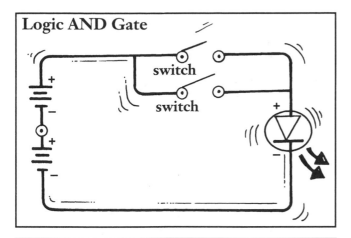

### Logic OR Gate

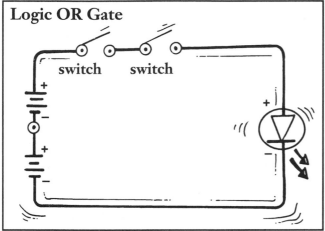

these components will not work if they are installed incorrectly.

The circuit in which the two switches are in parallel with each other represents the OR computer logic gate. If either the first switch *or* the second switch is placed in the "on" position, the LED will light.

The circuit in which the two switches are wired in series with each other represents the AND computer logic gate. Both the first switch *and* the second switch must be on before the LED will light.

Compare all possible combinations of on/off positions and the switches with the AND and OR logic tables shown in the chart. Replace each zero with the word "off" and each number 1 with the word "on." Try each switch combination as shown on the chart, placing the switches in various on and off positions and making note of whether the LEDs are lit or not.

tion, either the first fence gate or the second gate can be open to let cows pass.

Hypothesize that you can use toggle switches to represent the concept of AND and OR computer logic gates.

This project will require a little soldering. If you have never not done this before, ask someone experienced with soldering to show you the proper technique. You can practice by twisting two pieces of bare hookup wire together and soldering them. CAUTION: A soldering pencil tip gets quite hot. Use a stand to safely hold the hot soldering tool when you aren't actually using it. A piece of cardboard or wood under your work will catch any solder that drips. Be alert that the tip of the soldering tool can burn you. Have an adult supervise. As soon as you are finished using the tool, unplug it, but still be careful of the tip. Although it may not look hot, it will stay hot for some minutes after you disconnect it from the outlet.

Construct two circuits, one in which two switches are placed in series and the other in which two switches are placed in parallel. Be sure to observe the correct polarity when wiring the batteries and the LEDs into the circuit, since

### Gate Logic Tables

#### AND

| Input A | Input B | Output |
|---------|---------|--------|
| 0 | 0 | 0 |
| 0 | 1 | 0 |
| 1 | 0 | 0 |
| 1 | 1 | 1 |

#### OR

| Input A | Input B | Output |
|---------|---------|--------|
| 0 | 0 | 0 |
| 0 | 1 | 1 |
| 1 | 0 | 1 |
| 1 | 1 | 1 |

### Something more...

1. Some computer logic AND gates have more than two inputs. How would the above AND switch circuit be modified to simulate a 4-input computer logic AND gate?

2. There are other logic gates used in electronics: NAND, NOR and XOR. Research the logic tables for them. Can similar circuits using switches, batteries and LEDs be used to illustrate the concept of these gates?

# PROJECT 7–5
# Repulsive Attraction
## *Using sound waves to repel insect pests*

| YOU NEED |
| --- |
| • tape recorder and tape |
| • electronic keyboard or organ |
| • flashlight |
| • medium-size cardboard box |
| • piece of screen (large enough to cover the top of the cardboard box) |
| • a warm evening, when nocturnal insects are likely to be flying |

During the spring, summer, and autumn months, many of us are bothered by insect pests. Some insects, such as gnats, are so tiny that even well-screened doors and windows don't keep them out. Do nocturnal insects respond to sound? Are they either attracted to the source of it or repelled by it? Hypothesize whether or not you think sound has any effect on common flying nocturnal insects in your area.

Leave the first 15 minutes of an audio tape blank, with no sound on it. Then record 10 minutes of a continuous, very-low note, sustaining the key on a electronic synthesizer or organ. Follow that with 10 minutes of a continuous high tone.

Open out or remove the top flaps from a cardboard box about two-feet square. On a warm evening after dark, when insects are out, place the box outside. Put a lit flashlight and the tape recorder, containing the tape of low and high notes, into the box. Turn the recorder on and place a piece of screen over the top of the box.

The light from the flashlight will attract all kinds of nocturnal insects. Many will gather on the screen, where they can be roughly counted. The first 15 minutes of the tape has no sound, therefore, the light should be the only thing that is attracting the insects. Towards the end of the no-sound period, count the number of insects on the screen, doing it as accurately as you can. As the tape continues to play, the ten minutes of the low note will sound. After several minutes into it, count the number of insects now on the screen. Several minutes after the high note starts to play, count the number of insects on the screen again.

Examine your recorded data and reach a conclusion about your hypothesis.

**Something more…**
1. Does volume have an effect?
2. Use smaller incremental changes in frequency, perhaps recording five minutes of each key played on the organ or keyboard in an attempt to find a frequency that either attracts or repels insects.
3. Can you identify any of the type of species of insects that have gathered on the screen (beetles, butterflies, mosquitos, and so on)?
4. Instead of placing a tape recorder in the box, play a radio tuned to a station playing music. This would generate a wide range of frequencies at once.

# SECTION EIGHT
# RADIO-FREQUENCY ENERGY

Radio frequency waves, commonly called RF energy, are electromagnetic waves resembling light waves in their behavior. They travel at the speed of light. Radio waves are produced by an electronic high-frequency oscillator, a circuit that causes voltage to swing back and forth many times a second. This vibrating rate (the number of cycles per second) is called the "frequency" of the wave. There are various frequency "bands." These bands, or groups of frequencies, are used for different types of communications. On your radio, you may recall seeing the number KHz at one end of your AM dial (KHz means one thousand "Hertz" or cycles per second). Radio waves in the "AM broadcast band" are called "long waves." The "short wave" band begins where the other leaves off and is commonly used for foreign radio broadcasts. FM radio and television stations broadcast at much higher frequency bands, up in the millions of cycles per second. Extremely high frequencies in the gigaHertz range (1 billion cycles per second) are called "microwaves." Home satellite TV dishes and microwave ovens work at these frequencies.

Radio waves can be used to carry sound and pictures through the air. At a radio station, the music and announcer's voice is electronically combined with a "carrier wave," which is a high frequency radio wave. This wave is then transmitted from the station's antenna and it travels many miles through the air until it reaches the receiving antenna on your radio. The circuitry inside the radio allows you to tune in one station at a time and separate the sound from the radio-wave carrier. The transmission of television pictures, although much more complex, is handled in a similar manner.

The carrier wave can be amplified, made much more powerful, to increase the distance it can cover. Listen to your AM radio during the day and during the night. Log the stations you hear. Do you receive some stations at night that you can't bring in during the day?

# PROJECT 8-1
# Interference: Foiled Again!
## Atmospheric radio-frequency noise generated by electrical devices around the home

### YOU NEED

- personal computer
- AM radio
- coat hanger
- 2-foot-long piece of aluminum foil
- adhesive tape

One of the problems with electronic appliances with circuitry that operates at very high speeds is that they radiate RF (radio frequency) energy. Such appliances usually have a type of filter system or shielding to suppress RF radiation, but some energy may leak out and can often be detected in close proximity to the appliance.

One such offending electronic device is the personal computer. Computers, especially the newer models, operate at very high speeds, in—the MegaHertz range (1 MegaHertz is 1 million cycles per second).

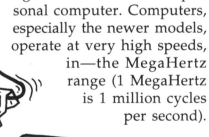

If radiation is too strong, this can pose a problem. If someone in your home is using a personal computer that is radiating a strong RF signal, it may interfere with someone else there trying to listen to a radio or watch television.

Hypothesize that radio-frequency interference generated by an electronic appliance can be reduced by the proper use of shielding.

Turn an AM radio on and set the tuning dial to a position where no station can be heard and the background noise is fairly quiet. Place a personal computer on a table or desk and turn it on. Bring the AM radio close to the computer. Do you hear a whistling sound from the radio? To be sure the RF signal being heard is radiating from the computer, turn the computer off. The whistling sound will disappear if its origin is the computer.

Using adhesive tape, tape a two-foot-long piece of aluminum foil to the bottom of a coat hanger. Holding the coat hanger, slowly pass the foil sheet between the computer and the radio. Keep the foil close to the radio to reduce the amount of RF signal that leaks around the foil. Observe and record any changes in sound coming from the radio. Reach a conclusion about your hypothesis.

### Something more...
1. What happens to the whistling sound heard in the AM radio as you do different functions on the computer, such as access a disk drive or press keys on the keyboard?
2. Is the interference stronger on one part of the AM band than the other? (Is it stronger at the lower or higher frequencies, or doesn't it make any difference?)
3. Could this setup be used as a metal detector?
4. Try using other materials to shield the radio-frequency interference, such as plastic, glass, and wood.

# PROJECT 8-2
# Only Way Out
## *Limiting the direction of radio-frequency radiation*

**W**hen RF (radio frequency) waves are transmitted by a walkie-talkie, the waves leave the tall rod antenna in an "omnidirectional" pattern, that is, they radiate out in all directions equally. This is similar to the waves that ripple out when a stone is dropped into a pond. What if we wanted to limit the direction of the energy's travel? Hypothesize that it is possible to limit the direction of a radio wave.

Record your voice on a portable tape recorder, talking for about ten minutes. You could read part of a book or the daily newspaper.

Tear the flaps off of a cardboard box about two to three feet square. Cover all the sides of the box, including the closed bottom, with aluminum foil, taping it to the box.

Lay the box on its side outside so that the open side, the top, faces out. Place the cassette recorder with the tape of your voice in the box. Near the recorder's speaker, place one of the walkie-talkies, to catch the sound. If your walkie-talkie has a high, telescoping antenna, reduce the size of it so that will fit inside the box. Use a rubber band to depress the walkie talkie's transmit button, holding it "on." Start the cassette recorder playing.

With the second-walkie talkie tuned to the channel the other walkie talkie is transmitting on, walk all around the box and listen for any changes in the sound you hear. Does the sound become weaker when you walk behind the box or along the sides that have the aluminum-foil shielding? Reach a conclusion about your hypothesis.

### Something more...
Stake out the strong reception area in front of the box. Measure the angle of the strong signal coming out of the box.

### YOU NEED

- pair of walkie-talkies
- portable cassette recorder
- roll of aluminum foil
- medium-size cardboard box
- adhesive tape
- rubber band

# PROJECT 8–3
# Out of Sight
*Affecting the range of radio waves*

## YOU NEED

- "toy" walkie-talkie set
- a friend
- a long, straight street or section of roadway, with buildings or trees alongside

A common question people ask when purchasing walkie-talkies is "What distance will they cover?" The stronger the RF energy generated by the transmitter, the farther the signal will be clearly heard.

Toy walkie-talkies often have very little power—on the order of about 100 milliwatts, or one tenth of watt. This strength is only good for reliable communications around your own home and to your neighbor next door. Large walkie-talkies may put out as much as 5 watts. How far will that signal travel? The answer depends on many factors. If you are on the beach or in an open field, a 5-watt walkie talkie's signal may travel several miles. But deep in the woods, a mountainous region, or a city with tall buildings, the presence of obstacles will severely reduce that range.

Hypothesize that obstacles such as buildings and trees reduce RF-signal strength and limit the range of citizen-band radio communications.

Locate an area with a long, unobstructed view but obstructions close by, for example, a long straight roadway, with several houses or buildings along it. An open field next to a heavily wooded area would also work for this experiment.

Using two inexpensive, toy walkie-talkies, have a friend walk down the street (safely on the sidewalk or shoulder of the road) as you talk back and forth to each other. When your voices become hard to hear, stop walking. Make a mental note of how each of you sound and how far away you are standing. Without moving any further away from each other, turn and walk to the same side until you can no longer see each other, because of trees or buildings in between. As you continue to talk, do you notice any change in the sound of each other's voices? Have they become less audible and less clear? Reach a conclusion about your hypothesis.

**Something more…**

1. Some more expensive walkie talkies have signal strength meters on them. Use the meter to quantitatively measure the amount of signal reduction caused by the interference of buildings.
2. Does weather affect the range of communications? Consider rain, fog, high humidity, hot and cold temperatures.

# SCHEMATIC SYMBOLS FOR COMMON ELECTRONIC COMPONENTS

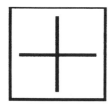

**wires intersecting (connected)**

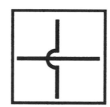

**wires intersecting (not connected)**

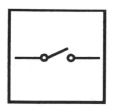

**switch**

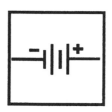

**battery**

**capacitor**

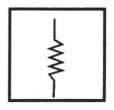

**resistor**

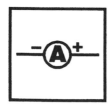

**ammeter**

**light bulb**

**diode**

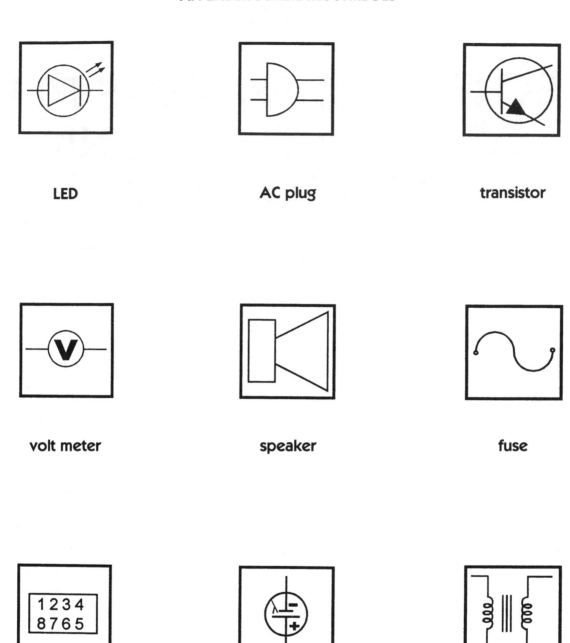

LED

AC plug

transistor

volt meter

speaker

fuse

integrated
circuit

solar cell

transformer

# GLOSSARY

**AC** alternating current, current that changes polarity (direction) rapidly.

**ampere** the unit of measure for current flow, often referred to simply as "amp."

**amplitude** the size or amount of wave form or signal. In electronics, often measured in volts.

**anode** the positive terminal on a battery or electronic component (such as a tube or a diode).

**armature** a rotating wire coil placed in a magnetic field. Armatures are found in generators and motors.

**capacitor** an electronic component that stores an electrical charge. It consists of two metal plates separated by an insulating material. The unit of measure of capacitance is the farad.

**cathode** the negative terminal on a battery or electronic component (such as a tube or diode).

**conductor** a material, typically a metal, through which electric current can flow easily.

**control** When doing experiments, a control has all the variables maintained. For example, to test the effects of carbon monoxide on plants, you must have two equally healthy plants. Both plants would receive exactly the same care and conditions (soil, sunlight, water). The experimental plant only would receive additional carbon monoxide. The other plant would be the control plant. The control plant would receive maintained conditions while the experimental plant receives the variation. Larger experiments often require a "control group."

**current flow** the transfer of energy along a conductor by the movement of electrons.

**cycle** Every time a voltage or signal reverses polarity and then changes back again constitutes one cycle. The polarity in the electrical outlets in the United States changes 60 times a second. Another term often used to describe a cycle is the "Hertz." One million cycles per second may also be referred to as 1 million Hertz, or 1 MegaHertz. The number of cycles per second is called "frequency."

**DC** direct current, unchanging current as supplied by a battery.

**dead short** an electrical path that has almost no resistance to impede current flow. This term is traditionally used to describe an electrical path of extremely low resistance (near zero ohms), where normally there would be a higher resistance. This signals a problem such as a failed electrical component or a faulty electrical path.

**electric circuit** various electrical components connected and arranged in a way to allow at least one closed path for current to flow.

**electrical potential** a measure of the ability of a battery or other power source to do work. Electrical potential is measured in volts.

**electrode** the terminal of an electrical source. The electrode with the positive charge is called the anode, and the electrode with the negative charge is the cathode.

**electrolyte** a liquid solution that conducts electricity and, when used with certain metals, can generate electricity. Typical electrolytes are a solution of salt water and a solution of diluted sulfuric acid.

**electromagnet** a magnet that only attracts when electric current is connected to a power source.

**electromagnetic field (EMF)** an induced field around a moving electric device or current through a conductor.

**electromotive force (EMF)** another term for "voltage," the electrical potential to do work.

**experiment** a planned operation designed to test a hypothesis.

**farad** the unit of measure of capacitance.

**flux** *See* **magnetic flux**.

**frequency** the number of times per second an alternating current completes a full cycle, going from a positive polarity to negative, and back to positive again.

**fuse** an electrical safety device which stops the flow of current in a circuit if the current flow increases beyond a desired level. Fuses are rated by the maximum number of amperes they will allow to flow before they "blow" and open the circuit.

**ground** electrical ground is a point in a circuit where there is a zero volt potential in reference to the Earth. The Earth always has a potential of zero volts.

**hertz** another term for "cycles," named in honor of Heinrich Rudolph Hertz (1857–1894) who discovered electromagnetic waves, the basis for radio and TV communications.

**hypothesis** a theory or educated guess. "I think, when asked how much they would weigh on Mars, more boys will have accurate guesses than girls."

**insulator** a material through which electricity will not flow.

**kilowatt-hour** the unit of measure of electrical power used by electric utility companies to determine how much electricity a customer uses. It is equivalent to 1,000 watts being used for 1 hour.

**load** the power absorbing device in a circuit.

**magnet** a material which exhibits the unique property of attracting pieces of iron and steel.

**magnetic flux** a field in the space surrounding a wire carrying an electric current or the field surrounding a magnet.

**micro** a prefix meaning one millionth. A microfarad is one millionth of a farad.

**milli** a prefix meaning one thousandth. A milliampere is one thousandth of an ampere.

**nano** a prefix meaning one billionth. A nanosecond is one billionth of a second.

**nichrome wire** an alloy of nickel, iron, and chromium which has a high electrical resistance as it heats and is used as a heating element in toasters.

**noise** random electrical voltages in the atmosphere and in electronic circuits which are usually undesirable. Noise may interfere with the operation of electronic equipment. Lightning causes electrical noise in the atmosphere which can be heard as sharp cracks in an AM radio.

**observation** looking carefully.

**ohm** the unit of measure of resistance.

**Ohm's law** Georg Simon Ohm discovered a relationship between voltage, current, and resistance in 1827, with the equation:
$$\text{current} \times \text{resistance} = \text{voltage}$$
where current is expressed in amperes, resistance in ohms, and voltage in volts.

**open circuit** an electrical circuit where there is a break in the path, preventing current from flowing.

**parallel circuit** an electrical circuit where components are connected across each other as opposed to connected end to end.

**pH** a measure of a liquid solution's alkalinity or acidity, neutral being a value of 7.

**photovoltaic cell** a semiconductor device that converts light into electricity (also referred to as a "solar cell").

**piezo electric** certain crystal materials, such as quartz and Rochelle salts, that generate electricity when their surface is subjected to physical force or sound pressure.

**polarity** the direction of current flow, which results in the condition of being positive or negative. One terminal on a battery is positive and the other is negative. Regarding magnets, the term polarity describes the property of the material having magnetic poles, one positive (attracting) and the other negative (repelling)

**potentiometer** a variable resistor, like the volume control on a TV or stereo.

**power** the rate at which energy is expanded. Power is measured in watts. One of the formulas for power is: watts = volts × amperes.

**quantify** to measure.

**RC circuit** resistive/capacitive circuit, an electrical circuit with a resistor and capacitor placed in series. When a resistor is placed in series between a capacitor and a power source, the time it takes to charge the capacitor increases as the value of the resistor increases.

**rectifier** a device that changes alternating current to direct current. A diode used to convert AC to DC is said to be acting as a rectifier.

**relay** a device in which one circuit can open or close a current path in another circuit. An electromagnet and spring are used to open and close switch contacts.

**resistor** an electronic component that impedes the flow of electrons. The unit of measure of resistance is the ohm.

**sample size** the number of items being tested. The larger the sample size, the more significant the results. Using only two plants to test a hypothesis that sugar added to water results in better growth would not yield a lot of confidence in the results. One plant might grow better simply because some plants just grow better than others.

**schematic diagram** a pictorial drawing of electronic components and how they are arranged and connected together in a circuit. Symbols are used to represent the components.

**scientific method** a step-by-step logical process for investigation. A problem is stated, a hypothesis is formed, an experiment is set up, data is gathered, and a conclusion is reached about the hypothesis based on the data gathered.

**semiconductor** a material that falls into a classification between conductors and insulators and exhibits unique electrical characteristics. Semiconductor components in electronics include diodes, transistors, and integrated circuits.

**series circuit** an electrical circuit in which there is only one current path and all components are connected end to end.

**short circuit** the falling of a circuit from its normal value to a very low value, which usually causes the electrical device to fail or a fuse to blow. Electricity takes a short cut through the circuit instead of flowing normally throughout all the components. If the resistance has become so low that it approaches zero ohms, it is said to be a "dead short.."

**sine wave** the fundamental wave form from which all others are derived. The amplitude (size) increases in one direction, then decreases and increases in the opposite direction, completing the cycle by returning back to the starting point. The swing of a simple pendulum traces a sine wave form. In the United States, the AC power delivered to homes by electric utility companies takes the form of a sine wave, with the voltage building positive then swinging negative 60 times per second.

**solar cell** *See* **photovoltaic cell**.

**static electricity** an electrical charge on an object giving it potential energy. The friction from rubbing a hard-rubber rod with fur causes electrons to be transferred from the piece of fur to the rod, giving the rod a negative charge and the fur a positive charge.

**thermistor** a device made of certain semiconductor material (such as uranium oxide, nickel-manganese oxide, and silver sulfide) which is temperature sensitive. As temperature increases, resistance decreases, behavior opposite to that of other metal conductors.

**tolerance-resistor band** The color of the fourth band on resistors indicates its "tolerance" or accuracy. A resistor marked 100 ohms with a silver fourth band (indicating 10% tolerance) could have an actual value between 90 ohms and 110 ohms.

**transformer** an electrical component consisting of two or more coils which can step voltage up or down.

**voltage** the potential difference in electric charge between two points. This potential differ-ence is the force which drives electric current through a conductor.

**watts** a unit of measure of electrical power.

**wave form** the shape of an electric signal.

**wet-cell battery** a container with two dissimilar metals immersed in a liquid solution causing a voltage potential to be produced across the two metals. Typically, the metals might be copper and zinc and the liquid a dilute solution of sulfuric acid.

---

## Metric Conversion Chart

### Inches to Centimeters
(12 inches=1 foot   3 feet=1 yard)

| | | | | | |
|---|---|---|---|---|---|
| 1/8 | 0.3 | 9 | 22.9 | 30 | 76.2 |
| 1/4 | 0.6 | 10 | 25.4 | 31 | 78.7 |
| 3/8 | 1.0 | 11 | 27.9 | 32 | 81.3 |
| 1/2 | 1.3 | 12 | 30.5 | 33 | 83.8 |
| 5/8 | 1.6 | 13 | 33.0 | 34 | 86.4 |
| 3/4 | 1.9 | 14 | 35.6 | 35 | 88.9 |
| 7/8 | 2.2 | 15 | 38.1 | 36 | 91.4 |
| 1 | 2.5 | 16 | 40.6 | 37 | 94.0 |
| 1 1/4 | 3.2 | 17 | 43.2 | 38 | 96.5 |
| 1 1/2 | 3.8 | 18 | 45.7 | 38 | 99.1 |
| 1 3/4 | 4.4 | 19 | 48.3 | 40 | 101.6 |
| 2 | 5.1 | 20 | 50.8 | 41 | 104.1 |
| 2 1/2 | 6.4 | 21 | 53.3 | 42 | 106.7 |
| 3 | 7.6 | 22 | 55.9 | 43 | 109.2 |
| 3 1/2 | 8.9 | 23 | 58.4 | 44 | 111.8 |
| 4 | 10.2 | 24 | 61.0 | 45 | 114.3 |
| 4 1/2 | 11.4 | 25 | 63.5 | 46 | 116.8 |
| 5 | 12.7 | 26 | 66.0 | 47 | 119.4 |
| 6 | 15.2 | 27 | 68.6 | 48 | 121.9 |
| 7 | 17.8 | 28 | 71.1 | 49 | 124.5 |
| 8 | 20.3 | 29 | 73.7 | 50 | 127.0 |

# INDEX

# INDEX

# INDEX

## About the Authors

ROBERT BONNET, who holds an M.A. degree in environmental education, has been teaching science at the junior-high-school level in Dennisville, New Jersey, for over twenty years. He was a State Naturalist at Belleplain State Forest in New Jersey and has organized and judged many science fairs at both the local and regional levels. Mr. Bonnet is currently chairman of the Science Curriculum Committee for the Dennisville school system and a science teaching fellow at Rowen College in New Jersey.

DAN KEEN holds an Associate in Science degree, having majored in electronic technology. A computer consultant, he has written many articles for computer magazines and trade journals since 1979. He is also the co-author of several computer-programming books. In 1986 and 1987, he taught computer science at Stockton State College in New Jersey. Mr. Keen's consulting work includes writing software for small businesses, and teaching adult-education classes on computers at several schools.

Together, Mr. Bonnet and Mr. Keen have had many articles and books published on a variety of science topics, including, in 1995, *Science Fair Projects: The Environment* by Sterling Publishing.